The Takeover

environmental
history
and the
american
south

The Takeover

CHICKEN FARMING AND THE
ROOTS OF AMERICAN AGRIBUSINESS

Monica R. Gisolfi

The University of Georgia Press
Athens

Parts of this book appeared previously as "Leaving the Farm to Save the Farm: Poultry Farmers, Contract Farming, and the Necessity of 'Public Work,' 1950–1970," by Monica R. Gisolfi, in *Migration and the Transformation of the Southern Workplace since 1945*, edited by Robert Cassanello and Colin J. Davis (Gainesville: University Press of Florida, 2010), 64–79, and are reprinted with the permission of the University Press of Florida; chapter 1 appeared, in a different form, as "From Crop Lien to Contract Farming: The Roots of Agribusiness in the American South, 1929–1939," by Monica Richmond Gisolfi, in *Agricultural History* 80, no. 2 (Spring 2006), 167–89.

© 2017 by the University of Georgia Press
Athens, Georgia 30602
www.ugapress.org
All rights reserved
Set in 10.5/13.5 Adobe Garamond Pro by Graphic Composition, Inc., Bogart, Georgia

Most University of Georgia Press titles are
available from popular e-book vendors.

Printed digitally

Library of Congress Cataloging-in-Publication Data
Names: Gisolfi, Monica R., author.
Title: The takeover : chicken farming and the roots of American agribusiness / Monica R. Gisolfi.
Description: Athens : University of Georgia Press, [2017] | Series: Environmental history and the American South | Includes bibliographical references and index.
Identifiers: LCCN 2016039446| ISBN 9780820335780 (hard bound : alk. paper) | ISBN 9780820349718 (pbk. : alk. paper) | ISBN 9780820349459 (e-book)
Subjects: LCSH: Poultry industry—Georgia—History. | Poultry industry—United States—History. | Agricultural industries—United States—History. | Agriculture—Economic aspects—United States—History.
Classification: LCC HD9437.U63 G445 2017 | DDC 338.1/76200973—dc23 LC record available at https://lccn.loc.gov/2016039446

To Peter A. Gisolfi

CONTENTS

FOREWORD

For almost a decade, I lived in Athens, Georgia, and taught at the University of Georgia, and during that time I became deeply interested in the environmental history of the American South. Much of that interest was academic and bookish, but because environmental literacy begins with the local, I also spent a lot of time interrogating the landscape around me. Several things piqued my interest, including how extensively reforested the Piedmont had become after its long and damaging experience with cotton culture, and how, without much effort, one could locate beneath that mantle of trees the ghosts of soil erosion past. But I also had a series of memorable brushes with one of the region's dominant enterprises—industrial poultry production—that arose as cotton was in retreat. Anyone paying attention to the landscape of northeast Georgia in the early twenty-first century cannot fail to sense the poultry industry's ubiquitous footprint.

My impressions came in pieces and took a while to cohere. One of my vivid early memories was of attending my new faculty orientation and asking a colleague sitting next to me what department he was joining. "Poultry Science," he said, to my amazement. Until that moment, I had no idea there was such a field, let alone a whole academic department devoted (according to its website) to providing "outstanding educational experiences for students and service to poultry producers, poultry related businesses and the general public through the discovery, verification and dissemination of relevant, science-based knowledge." UGA's cutting edge Poultry Science program arose explicitly to serve that state's industrial poultry producers.

Soon thereafter, we settled into a beautiful house in a tree-lined neighborhood whose only drawback, we discovered, was an occasional putrid smell that wafted from the poultry processing plant less than a mile to the north of us. When I drove my children to their preschool, we passed right by that processing plant, and on the other end of our journey we drove right past a chicken feed mill as well. There weren't any hatcheries along this particular route, but there were several in other parts of town that I soon noticed. Together, hatcheries, feed mills, and processing plants composed the key components, the holy trinity, in the industry's vertical integration. And always there were dead chickens squashed flat on the feeder roads, birds that had perished in vain attempts to escape their industrial fate. My younger son, mildly traumatized by this constant exposure to industrial chicken production, swore one morning that he would never eat chicken again. Then he paused, considering the gravity of his commitment: "except for chicken nuggets," he added, perhaps because he could not yet imagine how living birds could be rendered in nugget form. The South, like much of the rest of the nation, is populated with fast-food places to ply your children with chicken nuggets, just one body of evidence for the dramatic growth in poultry consumption in the United States since World War II. And that growth in consumption would not have been possible without poultry processing plants, feed mills, and hatcheries.

Then, when a bad knee ended my rec league basketball career, I began cycling around Athens's scrubby hinterland for my exercise. A century earlier, cotton fields would have engulfed me, but by the early twenty-first century only a few small farms hung on. On many farm properties were long, low-to-the-ground buildings that I soon realized were chicken houses. I was confounded not just by their ubiquity, but also by how many of the older houses, though still structurally solid and respectable in appearance, seemed to be no longer in use. Although I did not initially realize it, I was witnessing one of the great innovations of the modern poultry industry, the grow-out operation, which has been partly premised on a continual process of rapid chicken house obsolescence encouraged by the industry. With my introduction to the poultry house, my education in reading the southern landscape of industrial poultry production was complete.

A fair amount has been written about the rise of the modern poultry industry, particularly from labor and environmental standpoints. Monica Gisolfi considers both of these issues in *The Takeover*, her wonderful primer on the rise of the industry in Georgia. But Gisolfi's great achievement is her focus on farms and farmers, and the industry's impact on them, for, as she masterfully shows, many farmers were lured into the grow-out business by the promise—a false one, it turns out—of making good money while staying on the land. As cotton culture collapsed around them during the years before World War II, northeast Georgians, some of whom had experienced the harshness of the crop lien system and tenancy, welcomed cotton's demise and sought out other economic opportunities. What they did not realize was that industrial poultry production would reproduce many of the pathologies of the waning cotton economy.

Gisolfi's main argument in *The Takeover* is that industrial poultry production in northeast Georgia represented one of the first and most influential manifestations of agribusiness. Today a small number of large firms—Perdue, Tyson, Pilgrim's Pride—produce a large percentage of the chicken consumed in America and, increasingly, around the world. Industrial processes, scale, and corporate ownership partly define these large firms, but vertical integration has made poultry production the "quintessential agribusiness" according to Gisolfi. And in northeast Georgia the vertical integrator par excellence was Jesse Jewell. When Jewell inherited his father's feed business in the early years of the Great Depression, he hit on the idea of encouraging farmers to raise poultry in larger quantities, which, conveniently, would consume his feed. As his business developed, he began providing not only feed but also chicks from his hatchery, and then he bought back the fully raised chickens and slaughtered them in his processing plant. He also provided credit to poultry growers. By the early 1940s, just before America entered the war, Jewell had in place the fundaments of a vertically integrated poultry firm.

While it might be tempting to tell Jewell's story as one of successful business innovation, of individual initiative rewarded, Gisolfi clearly demonstrates that federal policies and subsidies were indispensable to the rise of the Georgia poultry industry. New Deal allotment payments designed to reduce the production of certain crops allowed northeast Georgia landowners to decrease their reliance on cotton and invest in poultry. As a result, as Gisolfi memorably puts it, poultry houses replaced tenant houses, and some of the displaced tenants migrated to town and secured poultry processing jobs. Then, during World War II and as a result of red meat rationing, the federal government actively encouraged industrial poultry production. Indeed, the army became a major customer during the war, and in doing so propelled the standardization that favored large-scale producers. Moreover, the federal government began funding poultry science during the war, a trend that continued into the postwar years at places such as the University of Georgia. The war also drew a lot of labor off of the farms of northeast Georgia, and when those lucky enough to survive the war returned to the area, they often embraced poultry production as a postwar panacea. Gisolfi thus shows that federal policies and federal actions were essential to the success of Jewell and other poultry integrators at a transformative moment in the history of the nation's food systems.

After the war, the full implications of this new system of concentrated poultry production came into view, the result of several additional corporate innovations pioneered by Jewell. The most important, according to Gisolfi, was the feed conversion contract, the legal agreement between integrators like Jewell and the small farmers who grew out hatchery chicks into "broilers," the new term for the standardized chickens produced by the industry. Although ostensibly designed to give farmers a modest profit on the chickens they raised, these contracts, and the rage for efficiency they embodied, trapped "growers" in a vicious cycle of debt. Gisolfi refers to the system that evolved as "quasi-vertical integration" because integrators left the process of growing out chickens—of efficiently converting Jewell's feed into chicken flesh—to farmers. From the integrator's standpoint, this was a brilliant arrangement, as small farmers assumed responsibility for the riskiest, costliest, and most land-intensive part of the poultry production process. Moreover, the mechanism of the feed conversion contract allowed the industry to insist that farmers continually invest and reinvest in state-of-the-art grow-out facilities; if they refused, poultry companies channeled their hatchery chicks and feed to growers that cooperated. And when farmers purchased modern poultry houses and the machinery that went with them, they almost always did so from the leading purveyor of this equipment in the region: Jesse Jewell. The result was that northeast Georgia poultry growers were unable to escape escalating investment and debt. In this perverse system, they were at once capitalists and wageworkers. Gisolfi estimates that, incredibly, roughly half the capital investment in the industry came from these growers, though they enjoyed very limited control over the means of production and almost none of the profit. Moreover, the feed conversion contracts essentially dictated a wage to growers, based on conversion of feed to meat, that correlated little with market prices for poultry. Some "growers" made so little money from their operations that they had to take second jobs on the side—often low-paying off-farm poultry industry jobs. Thus, the poultry business, whose initial promise was to allow farmers to escape the tenancy and indebtedness to furnishing merchants that had plagued the old cotton economy, ended up replicating those very dependencies. Gisolfi's explanation of these agribusiness innovations and their impacts on rural economies and societies is *The Takeover*'s signal achievement.

The Takeover is mostly about the social costs of the rise of poultry production, though Gisolfi does address, both directly and indirectly, the many environmental transformations and costs that came with the integration of the industry. And while her focus is on northeast Georgia, the story Gisolfi tells has much wider implications for farmers and rural life in America. It is a story that has been largely repeated on the Delmarva Peninsula and in the Ozarks, the two other major poultry growing regions of the United States, as well as in other sectors of the animal protein industry, and it is a story whose elements can be found throughout a modern agricultural economy in which farmers are increasingly reliant on large corporations for patented seeds, chemical fertilizers, and pesticides. This is a tale about the beginning of the end of independent farming, a story of the remarkable increases in agricultural productivity and the concomitant cheapening of the American food supply, both of which, though they have had their demonstrable consumer benefits, have also come at high social and environmental costs in rural America.

The Takeover helped me to fully comprehend my various brushes with the poultry industry of northeast Georgia—the publically subsidized poultry science, housed within a state university, that has propelled the industry forward; the hatcheries, feed mills, and processing and rendering plants that speak to the industry's rapid integration; and the chicken houses, working and abandoned, that, like the tenant houses that preceded them, became an architecture of dependency in the region over the past half century. Like any good history should, then, Monica Gisolfi's *The Takeover* helps us to understand the world around us, a brave new world of modern agricultural production that extends well beyond the rolling hills of the Georgia Upcountry.

Paul S. Sutter
Founding Series Editor

ACKNOWLEDGMENTS

I am grateful for the opportunity to acknowledge and thank the many people who have contributed to this book's completion. I have relied on the expertise of the librarians and archivists at the University of Georgia, Columbia University, the Library of Congress, the National Archives, and the National Museum of American History. I offer special thanks to Joe Schwartz at the National Archives and to Jill Severn, Sheryl B. Vogt, and Steven Brown at the University of Georgia Libraries.

Foundations and institutions generously supported my scholarly work. Foremost was Columbia University, where this book started as a dissertation. I am also thankful for the support of the Smithsonian's National Museum of American History and the Whiting Foundation. One could not ask for more rigorous, intelligent, and generous professors than the members of Columbia's history department. Betsy Blackmar, Barbara J. Fields, Eric Foner, and William Leach's expertise, advice, and knowledge improved this book. At the National Museum of American History, Pete Daniel provided keen and insightful criticism of the entire manuscript.

I am fortunate to have benefited from the expertise of the editorial staff at University of Georgia Press: Jon Davies, Jim Giesen, Mick Gusinde-Duffy, and Bethany Snead. Paul Sutter's criticism and careful reading of the manuscript helped to make this a better book. I am grateful to Joseph Dahm for his most careful copyediting.

Colleagues and friends provided crucial comments, criticism, and support: Merlin Chowkwanyun, Jim Downs, Reiko Hillyer, Lu Ann Jones, Faye Kimerling, Sara Kramer, Timothy Pigford, Claire Strom, and Jeffrey Trask. Adrienne Petty read the full manuscript; her expertise and knowledge of the topic proved indispensible. My students in my environmental history seminars at UNCW whose enthusiasm, excitement, and commitment to the study of history continue to inspire my work. Barbara J. Fields remains my teacher. Her scholarship influences my scholarship and teaching. Her integrity and compassion, the standard she sets for teaching, scholarship, and advising, and her commitment to the study of history push me to be a better scholar. I am indebted to her in countless ways.

This book is dedicated to my father, who has supported my work from the outset and has argued with me over ideas, interpretations, and research. From a young age, I learned to look from my father—by which I mean my father prodded me to observe and scrutinize buildings, landscapes, streetscapes, urban plans, and the way in which people interact with landscapes, the built environment, and nature. He taught me about the importance of place and pushed me to consider the relationship between landscape and history. My father's work ethic, integrity, and commitment to teaching, research, and design inspire my own research and teaching.

The Takeover

INTRODUCTION

If you fly into Atlanta's Hartsfield-Jackson Airport, you will most likely fly above the Georgia Upcountry, a region just north of the city. As you make your descent, you will see hundreds of nondescript but efficient-looking groups of structures dotting the landscape. Four long, uniform structures, barns of some kind, stand perfectly parallel to one another. Sitting adjacent to these generic buildings are a feed silo, generators, and massive brown pools of dense waste. To most Americans, these structures will appear unfamiliar, if not unrecognizable. This is the American farm. Your skinless chicken breasts, your rotisserie chicken, your preseasoned and cooked Perdue chicken strips, your chicken tenders, your Chicken McNuggets, your Chick-fil-A sandwich all come from these "farms" to your table.

Few visitors to Atlanta venture beyond the metropolitan area. But if you were to visit the Georgia Upcountry, you would drive north from the city about sixty miles on the interstate. As you pass the structures that make one place in America indistinguishable from another—the big-box retailers, office parks, megachurches, and gated communities—you will start to see a change in the terrain. The air becomes a bit cooler, and gashes appear in the earth through which the red soil bleeds. When the terrain becomes hilly, you have reached the Upcountry, "the red hills of Georgia." Here, you will see abandoned farmhouses and churches that will remind you of Walker Evans's stark photographs of the Depression-era South. Every so often you will find evidence that Upcountry farmers, as recently as fifty years ago, planted cotton as their sole cash crop: single cotton stalks poking through otherwise abandoned fields, empty and dilapidated cotton storage warehouses. The scene is picturesque, pastoral—a tableau that seems sculpted from frozen time.

This would-be pastoral, rolling countryside is devoid of tractors, animals, and crops. Nevertheless you are now traveling through an important part of Georgia's very profitable poultry industry. A major component of global industrial agriculture is hidden in the midst of these rolling hills. Farmers raise record numbers of chickens here every year. The scale of this activity is anchored not to any indigenously shaped rhythm of life or memory-bearing landscape but to the global industrialization of food production.

It is also a landscape remarkably less populated but far more polluted than in the past. In 1993 the U.S. Census Bureau ceased counting the number of Americans living on farms.[1] Roughly fifteen years later, the Environmental Protection Agency reported agricultural runoff as "the single largest source of water pollution in the nation's rivers and streams." In 2005, one state's attorney general sued thirteen poultry companies, finding that "they had damaged one of the state's most important watersheds."[2] According to the U.S. Department of Agriculture, "The average American is exposed to 10 or more pesticides every day, via diet and drinking water."[3] Taken together, these data demonstrate the scope and consequences of capital-intensive industrialized agriculture, better know as agribusiness.

The chapters that follow examine the growth of the poultry industry in Upcountry Georgia, the self-proclaimed "poultry capital of the world," focusing on the four counties where the industry coalesced: Cherokee, Forsyth, Hall, and Jackson. This region became home to the quintessential agribusiness—commonly understood as a system of capital-intensive industrial agriculture—resulting in environmental degradation and a class structure that reduplicated the labor exploitation of an earlier era. This study is neither a comprehensive history of the American poultry industry nor a comprehensive history of American agribusiness. Rather, it examines a region that originated the silent revolution that unfolded over the course of the twentieth century: the shift from small farming to capital-intensive industrial agriculture. Thus, though this is a book about a small place, it is by no means a small story.[4]

Well before the Great Depression, feed merchants in the Upcountry, with the backing of national feed manufacturers, adapted the crop lien system—the oppressive farm credit system that had facilitated cotton production since the end of the Civil War—to the production of chicken. In the process, cotton farmers, most familiar with the crop lien system, became poultry growers. Government subsidies played a crucial role, driving the growth of the industry, often if not always to the detriment of small farmers. Farmers from the 1950s onward acquiesced to industry demands, assuming sizeable debt to provide more than half of the capital to run what has now become a billion-dollar industry. This growing industry, over time, created massive amounts of animal waste, polluting water, air, and soil. Slowly this wildly successful business model, known as quasi-vertical integration, became a blueprint that shaped the rise of modern agribusiness, both nationally and internationally.

The story begins in the Upcountry, where, from Reconstruction onward, cotton farmers and furnishing merchants utilized the crop lien system to finance, control, and market the crop. Merchants and landlords, charging high interest rates, advanced credit to farmers in the form of seeds and fertilizer.

At settlement time in autumn, after the cotton was picked and baled, farmers paid their creditors. During the first decades of the twentieth century, the cost of production persistently exceeded the market price of cotton. Farmers began to assume crushing debt. This system, in part, contributed to the rampant poverty that plagued the South well before the Great Depression. By the 1930s, prolonged agricultural depression and New Deal reforms that restricted cotton production made cotton farming untenable in the Upcountry. Feed merchants in Upcountry Georgia, with the aid of national feed manufacturers, became industry leaders by adapting the crop lien system to the production of chicken. In the process, cotton farmers became poultry growers, who from the 1950s onward acquiesced to industry demands, taking on sizeable debt much in the same way that their forebears had acquiesced to the demands of cotton furnishing merchants. By the 1950s, poultry had become Georgia's most important farm product. Over time, farmers found that they had replaced one form of one-crop agriculture with another, and one form of debt for another, helping to create a system that reinforced the widespread poverty and intractable class stratification long associated with cotton planting.

This study is primarily, if not wholly, the story of white businessmen, USDA bureaucrats, and white farmers holding small plots. Integrators and the federal government actively and intentionally pushed African Americans and women to the margins of this system. Yielding to pressure from integrators, white farm women ceased raising seasonal flocks. As the USDA introduced the newest technologies to white farmers, it simultaneously pushed African American farmers to raise solely yard flocks, consigning them to operations too small to rival those of white farmers. The fact that adult black farmers learned the very same curriculum white children were learning in their 4-H clubs reveals the unequal access to information that characterized agriculture in the Jim Crow South and the limits to African American participation in the poultry industry. The absence of African Americans in the chapters that follow is a function of racist USDA policies that kept them from benefitting from the research, technology, and subsidies poured into the development of this industry.[5]

Chapter 1 explains the process by which local merchants, national feed companies, and the federal government encouraged landed farmers to shift from cotton to chicken production. In so doing, they adapted the exploitative crop lien system to raising poultry, thus laying the foundation for the modern poultry industry by the end of the New Deal. Chapter 2 examines food policy during World War II and shows the federal government's crucial role in developing the poultry industry. Far from being an inevitable result of innovative farmers and technological improvements, the transformation of chicken production was driven by war food policy, government regulations, and government-funded research. In the third chapter, I explain the business model developed by the leaders of the poultry industry, known as integrators. The archetype for the modern poultry industry, this business model was both highly profitable for integrators and highly exploitative. In chapter 4, "Broiler Sharecroppers and Hired Hands," I look at the human cost of the business arrangement developed by integrators. Heads of the poultry industry instituted contract farming and demanded that farmers increase their investments in chicken machinery and housing. Poultry farmers, though nominally independent, became contract farmers for large poultry firms, which dictated how they raised chickens, required that they buy expensive equipment, and demanded that they furnish roughly half the capital required by what was becoming a multibillion-dollar industry. Farmers moved between farmwork and wage work to finance industry growth and to supplement their declining farm incomes. The final chapter examines the environmental consequences of industrial poultry production. Furthermore, it explains how the citizens of Upcountry Georgia fought against these consequences and the limits to their success.

The birth and rise of the Upcountry's poultry industry warrants investigation. This story illuminates the evolution of agribusiness. It explores the rampant inequalities central to its growth and examines how food production contaminates the air we breathe and the water we drink. Here in the Upcountry we can view the origins of the silent agricultural revolution that has forever changed the United States.

CHAPTER I

From Cotton to Chicken, 1914–1939

IN 1914, a Georgia businessman painted a bleak portrait of the results of Georgia farmers' heavy reliance on cotton cultivation. He complained that one could ride a hundred miles without seeing a herd of livestock and added, "When you do see cattle they are little tick infested creatures that no more resemble real cows than a tubercular cotton factory operator resembles an athlete."[1] From the end of the Civil War through World War II, farmers in Upcountry Georgia grew record amounts of cotton, and little else, on increasingly infertile land. In the first three decades of the twentieth century, cotton production continued and conditions among farmers grew worse, as they lost their land and slipped into the ranks of tenant farmers and sharecroppers. A region once inhabited by small landowning farmers became a land dominated by furnishing merchants and cotton farming, plagued by soil erosion, tenancy, and sharecropping.[2] Cotton production had obstructed diversification and left Georgia farmers impoverished. "There is no grain, no hay, no poultry, no vegetable gardens, no orchards—except the peach orchards belonging to non-resident corporations," a businessman observed, "nothing that goes to make up a real farmer's home."[3] A Georgia farmer reiterated the concerns of that Georgia businessman, insisting, "What is hurting poor people here the worst of anything is trying to raise cotton to buy supplies with." If farmers diversified, produced a range of food crops in addition to cotton, the farmer reasoned, Georgia farmers and their families would be "the happiest people in the world."[4] Andrew M. Soule, a leader of the new and expanding state agricultural bureaucracy and the president of the Georgia State College of Agriculture, disparaged the Georgia farmer who, in his view, devoted "so much of his time and attention to the cultivation of a single crop." "Is it necessary," he asked in 1923, "that we should adhere to all-cotton in Georgia, as many of our people think? No, not at all. It should be our surplus money crop—our servant, and not our master."[5]

In the first decades of the twentieth century, landowning farmers began the protracted process of shifting from cotton farming to chicken raising. Local merchants, national feed companies, and the federal government led the way, encouraging landed farmers to enter the chicken business. These four groups laid the foundation for a poultry industry that would boom during World War II and would soon govern the economy, the land, and the people of Upcountry Georgia.

In the 1930s, proprietors of country stores, known as furnishing merchants, adapted the cotton crop lien system to poultry raising. To be sure, cotton farming by no means disappeared between World War I and World War II. Not until the end of World War II could Upcountry farmers safely and triumphantly say they were no longer under cotton's yoke.

Part 1. The Crop Lien and the Persistence of Cotton

Cotton and its companion, the crop lien, had not always dominated Upcountry Georgia. Prior to the Civil War, farmers in the region produced little to no cotton on diversified farms. Making up 25 percent of Georgia's white population, farmers in the Upcountry produced less than 10 percent of the state's cotton before the Civil War.[6] In an area dominated by white landowning farmers on small farms, farmers generally grew enough grain and meat to feed their families and rarely relied on country stores for essential food supplies. Until the Civil War, Upcountry farmers "remained on the periphery of the export economy," according to Steven Hahn, their principal historian.[7] However, the war devastated Upcountry Georgia. When Civil War veterans returned home, they found weed-filled fields, broken fences, and empty storehouses. Confederate soldiers seeking provisions, followed by General William Tecumseh Sherman's troops, stole livestock and destroyed crops.

Postwar conditions chipped away at the self-sufficiency that had once characterized life in the Upcountry. In the years following the war, as railroads tied the region to northern markets and later to international markets, Upcountry farmers increasingly turned to merchants for the supplies needed to plant their crops. Storekeepers, known as furnishing merchants, began to demand a lien or mortgage on the crop in exchange for credit, and required that farmers grow cotton, which could be sold in northern markets. The cotton crop served as collateral. Forced to produce cotton to secure credit, farmers in Upcountry Georgia began reducing corn, wheat, and other foodstuff acreage and increasingly relied on merchants not only for fertilizer and cottonseed but also for bacon, corn, and other necessities.[8]

Upcountry farmers did not choose to grow cotton year after year; their creditors demanded it. Hahn explains that the "new and exploitative credit system . . . tied smallholders firmly to staple agriculture," and by 1890 "the Upcountry stood fully transformed, wrenched from the margin into the mainstream of the cotton market."[9] As furnishing merchants advanced farmers seed, fertilizer, and other necessities on credit, they refused to offer credit on any crop but cotton. Likewise, landed farmers often owned commissary stores where they sold foodstuffs to their tenants and sharecroppers and earned handsome profits. Merchants depended on the sale of foodstuffs to farmers and therefore had little to gain and much to lose if farmers could raise food for themselves. The interests of farmers directly competed with those of creditors.

In the first three decades of the twentieth century, key parts of the Upcountry's farm economy remained relatively unchanged. Rates of tenancy, average farm size, acres of cotton planted, and bales of cotton produced remained almost constant.[10] Likewise, the total combined population in Cherokee, Forsyth, Hall, and Jackson counties hovered around eighty-three thousand between 1910 and 1930. The number of farms did begin a slow decline between 1910 and 1930.[11] County agents, those charged with aiding landed farmers, lamented that little changed during these decades. Their reports tell of a stagnant and impoverished region. The county agents found it impossible to create diversified farms, and often found that their "scientific agriculture" fell on deaf ears. Between 1900 and 1930, the number of tenants in the four-county area continued to exceed the number of landed farm operators. Roughly 60 percent of those who tilled the soil were tenant farmers.[12]

In the Upcountry in those three decades, more remained the same than changed. During the roaring twenties, when many Americans purchased consumer goods, Upcountry Georgians—like most rural people—sat on the sidelines, reading the Sears catalog by candle light, lacking money and electricity for the new world of goods: refrigerators, toasters, washing machines, and so forth. Likewise, automobiles, tractors, and cotton pickers were almost entirely absent in the area. Fewer than 15 percent of southern farmers had a telephone, as compared to roughly 40 percent of the nation's farmers as a whole, and in 1920 a mere 1 percent of southern farmers owned tractors.[13]

It would be false to say that all remained stagnant in the Upcountry in these thirty years. Soil exhaustion increased, in part because the system of tenancy impeded its conservation. It is estimated that 30 to 50 percent of the tenants and sharecroppers moved each year, constantly seeking a better deal or a more fertile plot. Few, if any, sharecroppers and tenants had the means or time to improve the soil.[14] A plague of insects compounded rural poverty, with the boll weevil devastating crops in the mid-1920s. One government study found that Georgia and South Carolina led the way in the decline in crop acreage. Between 1920 and 1925 roughly 3.5 million acres of arable land were removed from production in the state of Georgia.[15]

The price of cotton by no means remained stable, and glutted cotton markets only increased poverty. The exigencies of the international market devastated the Upcountry. From 1900 onward studies by the Georgia State College of Agriculture and the U.S. Department of Agriculture (USDA) reported that high cotton prices did not translate into prosperity for many cotton farmers. A 1913 report made clear that profits on cotton averaged less than a cent per pound. Even though cotton prices reached a record high during World War I, a 1918 survey determined that 44 percent of farms lost money on cotton despite the high prices, which could not compensate for the high costs of fertilizers and other goods bought from the furnishing merchant, who charged high interest rates.[16] In 1919, cotton sold for thirty-five cents a pound. The following year, the price of cotton dropped to sixteen cents a pound. Surveys conducted mid-decade concluded that the situation had only deteriorated. A 1925 study reported that the average Georgia cotton farmer lost nearly five dollars per acre.[17]

In the Upcountry, landowners, tenants, and sharecroppers continued to plant cotton because their creditors refused to extend credit on other crops. Cotton may have been to blame for the economic plight of Georgia farmers, but cotton plants did not arise each planting season, purchase costly fertilizers on credit, force themselves on Georgia farmers, and then march their way to already glutted markets. Georgia farmers, at the behest of their furnishing merchants, planted the cotton. In 1921, the Cotton Pageant at the Northeast Georgia Fair demonstrated that farmers had little love for their staple crop. An article in an Upcountry Georgia paper announced that "if you want to see the boll weevil get what's coming to him, don't miss the Cotton Pageant. . . . Prince Diversification gets on [the boll weevil's] trail and, after a spirited contest, lays him low." It is noteworthy that King Cotton was not the one to destroy the boll weevil; rather, it was "Prince Diversification" who would be teaching the boll weevil a lesson, in what the article promised to be not only an amusing display, but also a "highly gratifying" one.[18] In 1919, the residents of Enterprise, Alabama, literally placed the boll weevil on a pedestal, erecting a fourteen-foot-tall statue in its honor. The monument's boosters claimed that this unlikely hero had been the "Herald of Prosperity" and had brought diversification to Enterprise by proving the folly of a one-crop economy.[19]

Folk and blues music, popular in the South in the first decades of the twentieth century, highlighted the oppressive crop lien system. "Down on Penny's Farm," a popular song in the South in the early decades of the twentieth century, suggests that the crop lien plagued southern farmers who grew cash crops like cotton and tobacco. The song, narrated by a tobacco tenant farmer, explains the practices of a landlord named Penny.

It's hard times in the country,
Down on Penny's farm
Now you move out on Penny's farm,
Plant a little crop of 'bacco and a little crop of corn.
He'll come around to plan and plot
Till he gets himself a mortgage on everything you got.

You go to the fields and you work all day,
Till way after dark, but you get no pay.
Promise you meat or a little lard,
It's hell to be a renter on Penny's farm

This song of protest reveals the fears and anxieties of southern farmers. In its conclusion, the farmer "goes down in his pocket with a trembling hand," telling Penny that he "can't pay you all but [he'll] pay you what [he] can." Penny's tenant is then sent to the chain gang for failing to make payments.[20] Folk songs like "Down on Penny's Farm" and blues musicians who sang of the "hungry time" tell us—in a very intimate way—about the lives of farmers.

Without adequate financing, farmers simply could not raise other crops, and thus poverty persisted in Upcountry Georgia. The work of one county agent attests to the obstacles that farm families and reformers faced. With the goal of teaching homemaking and improving the farm home, one home demonstration agent in Georgia sponsored a table setting contest for farm girls. One young girl entered the contest, knowing she could not win first prize because her family owned only one knife and one fork. Nonetheless, she took third prize, a set of silver teaspoons. According to the home demonstration agent, the young girl "had cut pictures of flat silver from magazines and had practiced laying the cover with these pictures." The agent celebrated the pluck of the young girl and boasted that she planned to sell canned products to buy a full set of silverware for her family.[21] To be sure, the family's ownership of just a knife and a fork suggests that their problems ran deeper than simply a lack of silverware. In Upcountry Georgia, home demonstration agents sought to aid Georgians who lived in dire poverty. Yet agents were often equipped with the most inadequate of resources to tackle deeply rooted economic problems.

As the country entered the Great Depression, few Upcountry Georgians could imagine an end to cotton farming. Cotton had dictated the rhythm of their lives and dominated the area for decades. That would soon change. Cotton prices fell in the first years of the 1930s, and Upcountry farmers fell deep into poverty. In the state as a whole, the gross income for farmers dropped from $206 in 1929 to $83 in 1932.[22]

Lean year followed lean year. Merchants went bankrupt and banks failed, as low cotton prices prevented their clients from paying bills. One study of the credit system in Georgia explained that, increasingly, farmers were unable to pay debts in poor crop years. The logic of the lien entailed that a year of loss would be followed by a profitable year, allowing creditors to collect old debts. However, in the 1920s, creditors found it increasingly difficult to collect from borrowers. This inability, in the words of one survey, impaired "the lending power of local banks and credit merchants," leading to "a general restriction of credit."[23]

Writing in 1929, rural sociologist Rupert Vance argued, "There exists a fundamental conflict between the public needs of the region and the vested interests of those engaged in supplying the various forms of cotton credit."[24] Cotton farmers could not change the system. They could neither opt to plant sweet potatoes and corn nor expand their flock of laying hens and spring chickens and expect to receive goods on credit. Consequently, they continued to plant cotton, selling it for less than the cost of production. Vance argued that change would have "to be engineered from above by bankers, landlords, and supply merchants." Those in power would have to undergo "a social crisis such as a continued depression in the cotton market." Only after a period of "continued loss," he reasoned, would bankers, merchants, and landlords allow farmers to cease cultivating cotton.[25]

With the onset of the New Deal, the federal government came to the countryside to control cotton production. New Deal policies placed restraints on cotton farmers but did not deal cotton a death blow. It would take World War II to bring an end to cotton production and tenancy in the Upcountry. During the war cotton production fell precipitously, as farmers in the four-county area, who had produced 65,484 bales of cotton in 1929, produced just over 29,000 bales in 1945 and fewer than 15,000 in 1949.[26]

Part 2. Uncle Sam Comes to Town

In 1933 the federal government intervened to repair a broken agricultural system. In late summer that year, at the behest of the newly formed Agricultural Adjustment Administration (AAA), farmers measured their crops and plowed up a third of their cotton. The plow-up of 1933 has been remembered as a dramatic moment, but it was far more than this. The federal government, specifically the USDA, intervened directly in the economy and in the very choices farmers had once made. This was the moment when the USDA began the process of supporting large farmers at the expense of small farmers and dismantling if not destroying small farms. Those who were first experiencing the New Deal farm programs, however, had no way of predicting their long-term consequences.

From 1933 onward, cotton farmers took acreage out of production, requiring smaller amounts of cottonseed and fertilizer than in years past. Vance's prediction—that cotton cultivation in the South would decline only when merchants and landlords had undergone a period of "continued loss" that was "painful and of long duration"—slowly became reality. Seeking to control crop surpluses that were glutting markets and driving down prices, the government introduced allotments. The government paid landed farmers to take land out of production, effectively "renting" land that lay fallow. Throughout the South, many, if not most, landed farmers pocketed the allotment money, sent tenants out into the fields to plow up a third of the crop, and then cast them off the land or hired them as day laborers. In the state as a whole, ninety-seven thousand farmers plowed up seven hundred thousand acres of cotton, for which the government compensated them roughly $8 million.[27]

Farmers and AAA agents exchanged words and sometimes blows over allotments, measurements, and the fact that the federal government was taking a central role in the lives of farmers. As a young man, president-to-be Jimmy Carter measured cropland for the AAA in southeast Georgia to determine if farmers were planting their allotted acreage. "Meticulous in [his] work," Carter nonetheless found that his calculations sometimes contradicted those of landowners, a discrepancy that could erupt in violence. Carter remembered one farmer who stormed into the Carter family store, grabbed him, and shouted, "Why the hell are you trying to cheat me out of my government payments?" When Carter attempted to explain his calculations, the farmer remained unconvinced, threw Carter to the ground, and began punching him.[28]

Conflict and resistance to government intervention pervaded the South. A *New York Times* reporter, Charles Puckette, headed south in the summer of 1933 to investigate the plow-up and found opposition to the measure. One important group that deserved mention, in the journalist's view, was an essential part of the South's workforce: mules. "The mule has been trained to walk between the rows and not to tread on the cotton plants," Puckette explained. However, New Dealers "asked this conservative to change his ways, to trample on the rows as he dragged the destroying plow."[29] Farmers balked just as hard at the thought of intentionally destroying fields of cotton. It just seemed "wrong before God to plow up a crop," admitted one. A friend of the Carter family remembered that he "couldn't keep [his] mule up on the row, where she had never been before without being whipped." He explained, "I had to let her walk near the middle, and hold the plow way over sidewise to reach the cotton stalks," adding, "It was hard work, and I almost cried." In Georgia's Greene County, just south of the Upcountry, sociologist Arthur Raper overheard tenants discussing the plow-up. One remarked, "I ain't never pulled up no cotton stalks befo', and somehow I don't like the idea." Another tenant proposed the following: "Let's swap work that day; you plow up mine, and I'll plow up yours." In fact, swapping became common practice throughout the South, where farmers could not bear the thought of plowing up their own rows of cotton, destroying the fruits of their labor.[30]

Despite human and animal opposition, federal support for reduced production continued. In April 1934, Congress passed the Bankhead Cotton Control Act to thwart the efforts of farmers who evaded allotments. The act set marketing quotas, limiting the amount of cotton that farmers could sell. In 1936, when the Supreme Court invalidated key provisions of the AAA, the Roosevelt administration fired back with the Soil Conservation and Domestic Allotment Act, which paid farmers for shifting land from cotton to soil-conserving crops. In Georgia cotton cultivation fell from 3.4 million acres in 1929 to under 2.2 million acres in 1939. Cotton farmers—both those with land and those without—began plowing up portions of their crops in return for government payments that went directly into the pockets of landed farmers, not those of tenants and sharecroppers. Landed farmers soon bought fewer and fewer sacks of cottonseed and fertilizer, and furnishing merchants found once-bustling stores as empty as the surrounding cotton fields.[31]

The drama surrounding the plow-up and subsequent government policy obscures the fact that New Deal reforms reinforced the existing power structure in Georgia and the South. President Roosevelt, in return for the white southern vote, stepped aside and allowed the white elite to oversee agricultural policy. The county elite sat on the committees that oversaw the distribution of AAA allotments and administered the AAA policies to their advantage, unremittingly strengthening their power, at the expense of black and white tenant farmers.[32] Payments intended for tenants and sharecroppers found their way back to landlords. One report explained the "wholesale neglect of the tenant" by pointing to the fact that the AAA "organized its program under the direction of the planters themselves." In his study of two Georgia counties, Arthur Raper explained, "Practically all of [the AAA money] found its way into the hands of the landlord. One-half of it belonged to him as rent, while the other half was used to reduce the tenants' indebtedness to him for furnishings."[33] The problem, the Hall County agricultural agent explained, was "getting the landlords to deal fairly with their tenants and share-croppers."[34] Landlords did not need to justify to tenants and croppers the taking of their payments. In the early stages, many tenants and sharecroppers were not even aware that they were entitled to payments, and those who were stood in no position to demand their checks from their landlords.[35] One study concluded that ultimately the AAA served as "merely a subsidy to planters."[36]

Not only did renters and croppers lose out on government payments, but small landowners lost as well. H. A. Maxey, the agricultural agent for Hall County, explained the plight of small farmers who, prior to New Deal reforms, produced "just enough cotton to supply their necessary things of life." When small farmers were forced to reduce their cotton acreage, Maxey stated, "it dug into their very existence." A large number of small landholders lived in the Upcountry, and they opposed the allotment program and crop reductions. According to Upcountry cotton farmer Guy Castleberry, large farmers were fine, but small farmers suffered: "The man like me that just had a few acreage, he just had to get out of it." For small farmers, taking land out of production left them with minimal acreage on which to grow a cotton crop. One county agent pointed out that small farmers could not reduce acreage planted in cotton without "reducing their standard of living which in many cases is very low now and has been for many years."[37]

AAA policies hurt tenants and sharecroppers. Landowners cast them off the land, as they removed acreage from production. Some allowed their tenants to stay, but benefit payments encouraged landowners to demote croppers and tenants to wage hands. Raper observed that AAA allotments "made it advantageous for the landlord to use wage hands instead of croppers." Census figures from Upcountry Georgia's Hall County testify to these practices. In 1930, 924 landowners and 1,805 tenants farmed in Hall County. By 1940 the number of tenants in Hall County had fallen to 1,351.[38] A handful of the 454 tenants who left the ranks secured land, but the vast majority became wage laborers or left the land altogether. In Cherokee, Forsyth, Hall, and Jackson counties combined, the number of tenant farmers dropped from 6,988 in 1935 to 5,366 in 1940. In that five-year stretch roughly 1,600 tenant farmers and their families left the land. This meant that close to one in four tenants farmers ceased farming.[39]

Landowners pocketed allotment checks, quite obviously a form of direct relief, and simultaneously voiced fierce opposition to relief measures for croppers and tenants, usually in the form of food and other basic necessities. Government handouts, landowners argued, would make tenants and sharecroppers unwilling to work for their keep. Landlords opposed relief measures for landless farmers because sharecroppers' and tenants' dependence on landlords for work made for loyal and docile workers. Relief measures, landlords claimed, would make already lazy tenants and sharecroppers even more shiftless. One report explained, "There is a considerable feeling among landlords that anything which disturbs this dependent status of the cropper is undesirable."

Landlords explained their opposition to relief in moral terms; they feared the "demoralizing effect" of relief on farm laborers. Few landowners voiced fears that government payments to the landed would have a demoralizing effect as well. After the season was over, however, landlords had no need for workers. Landed farmers then conveniently forgot their concern for the morality of their workers and eagerly endorsed federal relief measures.[40]

Some Upcountry farmers used government payments to improve farm buildings to house the Upcountry's new cash crop: chickens. With their allotment checks they expanded their chicken flocks, converted cotton-ginning houses to chicken houses, or tore down tenant houses, building chicken houses in their place. When they married in 1933, Tom and Velva Blackstock—both natives of the Upcountry—tore down a tenant house on their family's property and built a chicken house. Like many farmers who received allotment checks, the Blackstocks used the money to improve the farm buildings. Most likely, the tenant house they tore down had become vacant after the Blackstocks cast off their tenant farmers when the government began paying farmers to take land out of production.[41] As they expanded their flocks and built chicken houses, these farmers were laying the foundation for the poultry industry that would soon dominate the region and the economy.

Part 3. A Dumping Ground for Feed

In the early 1930s, Ralston Purina and other national feed manufacturers began developing a market for chicken feed in the South. Soon the region became a dumping ground for surplus chicken feed.[42] National feed manufacturers, one government report found, were the "biggest promoters of broiler production," and chicken feed accounted for most of the cost of production.[43] A government study of the Upcountry poultry industry attributed the rapid expansion of chicken production to the "the extensive use of credit" extended by national feed manufacturers.[44] Ralston Purina sales literature explained that it was difficult at first to sell feed to farmers who raised their own animal feed. "The selling of feed to farmers who raised their own grain and roughage for feed," the company literature explained, "was no easy task." The Purina Chow salesman had "a tremendous selling job to do." Local merchants became essential to the selling of feed. Merchants convinced farmers of the value of commercial feed and extended credit for baby chicks and commercial feed.[45]

Ralston Purina distributed feed and credit through local general stores and extended credit to southern merchants like Jesse D. Jewell—a native of Gainesville, Georgia—who would eventually establish the region's largest poultry firm, Poultryland Inc. Jewell's family owned a store in Gainesville that sold fertilizer and cottonseed meal and stored cotton.[46] Jewell's brother remembered that the family was "pretty well off in that day." Not only did the Jewells own horses, a buggy, four cows, two mules, and some chickens, but they were also among Gainesville's first families to own an automobile.[47] However prosperous the business may have been, it was not immune to the Great Depression. During the Depression, Jewell left college and returned home to assist his mother in the family store. The Great Depression devastated an already suffering region. Jewell recalled, "You just had to take a look at the unpainted farm homes, the eroded fields, the empty tables to know we were hurtin'." Customers were broke, and the family business, in Jewell's estimation, was "shot."[48] Cottonseed and fertilizer sales declined steadily, and excess animal feed sat in the family warehouse.[49] With the backing of national feed manufacturers, Jewell began advancing feed and chicks to farmers.[50] Studies of the poultry industry conducted in the 1950s confirmed that national feed manufacturers had been "particularly generous" in extending credit to furnishing merchants, in some cases requiring no collateral whatsoever.[51] Jewell promoted poultry raising. Soon residents of Gainesville and the surrounding areas followed his lead.[52]

The infusion of credit and feed began to change the Upcountry. Furnishing merchants developed chicken farming in the Upcountry as a market for surplus animal feed. The advent of cotton allotments left furnishing merchants with a Hobson's choice: they could develop new markets for products or go bankrupt. With the backing of national feed manufacturers, furnishing merchants shifted away from financing cotton production and began advancing farmers supplies to raise chicken. Furnishing merchants advanced farmers baby chicks and feed on credit, in much the same way that they had advanced farmers cottonseed, fertilizer, and other farm supplies. In 1935, Georgia farmers produced half a million chickens. Four years later, they produced three times that number.[53]

As Georgia farmers embraced poultry production, feed manufacturers and furnishing merchants reaped growing returns from chicken feed sales. Between 1935 and 1955, feed charges represented between 70 and 95 percent of the cost of producing chickens.[54] Most furnishing merchants were "primarily in business to sell feed, chicks, and other supplies needed by broiler growers," a government study later reported.[55] In 1936, J. S. Stephenson, a county agent, estimated that thousands of dollars of corn were being shipped into the Hall County each year. Two years later, Stephenson reported that farmers in Hall County purchased at least $150,000 worth of poultry feed—corn, wheat, and oats.[56]

When merchants began to promote poultry, Georgians already had extensive knowledge of raising chickens. For generations, chickens and eggs had been the domain of farm women and children, who tended yard flocks on farms throughout the South and traded eggs and chickens to supply their families' needs. Jimmy Carter, who grew up in south Georgia, recalled that his family "always had a yard full of chickens." A couple of times a year, Carter's father ordered from Sears and Roebuck hundreds of baby chicks, which the family ate and traded. While tenants and sharecroppers could not afford to purchase biddies in such bulk, they too had yard flocks. Ruby Faye Smith, the child of Upcountry tenant farmers, reminisced that her family "always had a yard flock," adding, chickens were "a part of life all the way."[57]

As long as general stores had existed in the Upcountry, farmers used chickens and eggs to barter for goods. "There wasn't any money along there from the late twenties on till about the late thirties," Spurgeon Welborn, an Upcountry farmer, recalled, so his family, like many others, swapped chickens and eggs for sugar, coffee, and flour. "Chicken eggs were a readily accepted form of currency," Jimmy Carter remembered, adding, "it was a matter of honor for a seller to assure their freshness, and there was an automatic replacement guarantee." Ruby Faye Smith's mother traded eggs and chickens for soap and coffee. Smith rebuffed the notion that the egg money belonged solely to her mother and was her "pin money." "There wasn't no 'my money and your money.' It was just lucky if we had money," she said.[58]

During the 1930s, women, who had traditionally managed the flock, began to earn rising profits. Describing herself as "not the clinging vine type," Mrs. O. H. Cooper expanded her flock and started raising chickens in 1930. "I get tired and discouraged, sure. No, everything don't run smoothly all the time by no means," she admitted in 1937, "but the joy and happiness of being the means of saving our home which we surely would have lost without my help, and my family having a decent living in the meantime, pays for all the discouragement."[59] In 1937, Miss Allie Street, who lived just north of Atlanta, raised 3,412 dozen eggs, earning $1,108.05.[60] In 1937, Mrs. Mark Davis, an Upcountry farmer, raised three hundred chickens. With her profits, Mrs. Davis bought a washing machine, refrigerator, and iron and electrified the family home.[61] Stories of chickens saving the farm abound. One farm woman paid off the loan on her family's farm with money raised from poultry production. Poultry allowed her and her husband to return to their farm. She explained that when the boll weevil hit, "March, 1926 we lost everything we had." She and her husband kept trying to grow cotton and began working for wages, but it was poultry raising that allowed them to return to the farm and make a living.[62]

When furnishing merchants began encouraging farmers to raise poultry for sale, many families initially expended very little money to enter the business. They drew on knowledge of poultry raising accrued over generations and expanded chicken coops, using material found on the farm to build chicken houses. Arthur Gannon, the state's poultry specialist, suggested that farmers convert old garages and vacant outbuildings to chicken houses.[63] Farmers re-used materials to build chicken houses, feeders, brooders, and waterers. One young boy gathered bricks in his pockets, "picking them up where he could find them discarded," and built a brooder, a device to heat the chicken coop. Arthur Flemming, an Upcountry farmer, reminisced that his family made their own feeders out of lumber found on the farm, built their own coal brooders, and mixed their own feed. When his grandfather purchased a watering device, Flemming recalled that it eased his workload a bit. When the device replaced the work of "going in there and filling up them little, old cans and things," Flemming recalled, "I thought we had it made."[64] Sanford Byers, another Up-country farmer, built a chicken house out of discarded lumber. He explained that he worked on the house when he "wasn't in a crop." Demonstrating thrift and ingenuity, Byers built a brooder out of an oil drum, and built feeders and watering devices.[65]

Dividing their time between row crops and their growing chicken flocks, Upcountry farmers finally saw their income rise. In 1936 Ruby Faye Smith's father, an Upcountry tenant farmer, entered poultry growing at the encouragement of his landlord and furnishing merchant. To Smith, the benefits of poultry growing seemed numerous: "When we got to raising chickens and got feed sacks, our problems were solved." Smith's mother used the feed sacks to make "panties, and slips, shirts, table cloths, sheets, pillow cases, dish towels." Her family was one of the lucky few that made the transition from tenancy to landownership, at a time when many tenant families were leaving farming altogether. Poultry promised a way for farmers to stay on the land, and the number of poultry farmers grew.[66]

In time chicken would become the main source of family income, pushing aside cotton, and women would cease being the lords of their flocks, but they by no means ceased working with chickens. As the industry grew, poultry did not shift from women's work to men's work. Men took control of decision making and women and children continued to work alongside their husbands, fathers, and brothers. "This poultry business got big," Upcountry farmer Spurgeon Welborn explained, and men in his estimation did most of the work. But the work of women and children remained indispensable, even as women lost control of the decision-making process. Historian Lu Ann Jones explains, "Women's loss of autonomy prefigured the erosion of independence that their men folks, in turn, would experience when they began growing broilers on contract."[67]

Part 4. Stabilizing Markets and Overcoming Nature

Access to ample credit drove the expansion of chicken farming, but as farmers expanded their flocks and sought to raise chickens year-round, they faced two key problems: seasonality and disease. For some time, farmers had primarily raised "spring chickens," which was a seasonal product. Seasonal production and "the spring chicken" inevitably created glutted markets in the spring and early summer. Clearly, farmers needed to find a way to raise chickens year-round. To produce year-round and especially in the winter months, farmers needed heated poultry houses and electricity. Government research and public works projects eventually enabled farmers to overcome the seasons, but in the early years county agents sought to manage glutted markets.

In the mid-1930s, the "spring chicken" led to a range of marketing problems. Farmers primarily sold their grown chickens in the spring and summer, but local markets were often glutted during this time; therefore, railroad car sales became essential. Often county agents oversaw and arranged these sales. On an appointed day and time, farmers loaded railroad cars with live poultry. "There was what they called a chicken car day," Upcountry farmer Spurgeon Welborn recalled, adding that these organized sales were "a big service because there wasn't any outlet for many chickens. You could flood the market or the grocery store with two or three families bringing what they didn't want," he explained.[68] As production increased, railcar buyers were unable to handle the growing number of chickens. Furthermore, some farmers found railcar sales costly because they had to travel over thirty miles to and from a sale.[69]

Farmers wanted a market "at their door," a county agent reported, to eliminate the cost and time of taking chickens to market.[70] They got their wish, and by 1937 poultry buyers were traveling through the Upcountry in trucks and purchasing live poultry directly from farms, saving farmers money and time.[71] Farmers sold their chickens directly to these buyers, who hauled chickens to the Atlanta market. In 1937, there were thirty different poultry truck routes in the Upcountry, and farmers sold more than 1.6 million pounds of chicken valued at almost $200,000.[72] As early as 1936, farmers in the Gainesville area supplied the entire Atlanta market in the spring.[73] However, as production increased, it soon became commonplace for the Atlanta market to be glutted in the spring. In some cases, county agents intervened again, contacting chicken dealers in Florida, who purchased Georgia chicken to sell in Tampa, St. Petersburg, Miami, West Palm Beach, Daytona, and Jacksonville.[74] To be sure, the "spring chicken" had become a marketing problem.

Markets were glutted in the spring and summer because farmers continued to raise chickens as a seasonal crop. "The majority of farm fryers," a Georgia poultry marketing specialist explained, were "produced and put on the market late in the spring and early summer."[75] The marketing specialist concluded that farmers needed to produce a constant supply of chickens "to maintain a regular market throughout the year." The specialist encouraged farmers to raise more chicken in the fall and winter.[76]

The New Deal's Rural Electrification Administration was indispensable to the growth of the Upcountry's poultry industry. Rural electrification made it possible for farmers to overcome the seasons and bring an end to the spring chicken. With electricity, farmers were able to raise chickens in confinement year-round, ending glutted markets in the spring. Intensive confinement was critical to the growth of poultry production, and government-funded research made year-round intensive confinement possible.

Electric power proved crucial to year-round operations. Before 1936, over 95 percent of rural homes in Georgia lacked electricity.[77] The Rural Electrification Administration changed that, and in 1936 over seven thousand rural homes in Georgia received electricity.[78] Lighting poultry houses proved key to increased production and raising chickens year-round.[79] Electric time clocks turned lights on and off and gave chickens "a working day of 12–14 hours," which meant that chickens slept less, ate more, and grew faster than they had without electricity.[80] Controlling light also meant that farmers could control the pace of the growing period: they could slow down the rate of growth if the market was glutted by keeping the lights on for fewer hours. Electric water pumps and electric waterers kept flocks supplied with fresh water.[81] Electric brooders—devices used to heat chicken houses—were controlled by a thermostat. Farmers thereby solved the problem caused by inconsistent heat sources—such as coal and oil brooders.[82] County agents reported that electric brooders were cheaper to operate than coal and brick brooders.[83] In 1938 T. J. Cox, the largest chicken farmer in the area surrounding Dalton, Georgia, installed electric devices that "lessened the burden of farm chores and made life 'more livable.'"[84] Electricity saved time and labor, enabling farmers to increase the number of houses and the number and size of their flocks.[85]

Over time, poultry production came to resemble industrial production. A government study found that "most of agriculture, unlike industry, is bound by the seasons rather than by the daily time clock." This was not so with poultry farming. The study explained that poultry production had come to resemble factory production; through government-funded efforts, the production cycle had been shortened, allowing farmers to raise multiple flocks. The study concluded that poultry farming was almost completely "divorced from seasonal influences."[86]

Once farmers heated chicken houses with electricity, they began intensive confinement—housing thousands of birds under one roof. Chickens no longer roamed the yard but now lived in chicken houses—hundreds and soon thousands tightly packed under one roof with no access to natural light. Significant problems accompanied this intensive, year-round confinement, namely nutritional deficiencies and infectious disease. Government-funded research helped to eradicate both.

Chickens raised in confined conditions, with limited access to ultraviolet light, suffered from a nutritional deficiency known as "leg weakness." Government research solved this problem. Scientists determined in the early 1920s that when added to chicken feed, vitamin D prevented this ailment. Cod-liver oil, rich in vitamin D, was henceforth added to poultry feed as a supplement. The discovery of vitamin D deficiency was a breakthrough. In the words of one environmental scholar, vitamin D "overcame one of the first major biological obstacles to industrial, continuous-flow production."[87]

Intensive confinement led to a range of contagious diseases—pullorum being the most devastating. In the 1920s and 1930s, pullorum destroyed entire flocks. The disease spread from infected hens to chicks, so eradication required the cooperation of hatcheries, farmers, and government scientists. The federal government helped to fight this contagious disease. It implemented the National Poultry Improvement Plan in 1935.[88] The plan reduced the incidences of pullorum and other diseases, established standards, and improved breeding qualities. Prior to 1935 there was no official testing for pullorum. Arthur Gannon, the USDA poultry extension agent for Georgia, observed that lack of testing was particularly detrimental to poultry production. He reported that before 1935 there was "no official blood testing in Georgia for pullorum disease and no system of approval or certification." Egg quality, Gannon maintained, was poor and "resulted in chicks lacking in size, livability and good breeding." According to Gannon, "The National Poultry Improvement Plan provides a set-up under which to cull breeding flocks and blood test them for pullorum disease under official supervision."[89] An official state agency—the Georgia Breed Improvement Supervisory Board—carried out the National Poultry Improvement Plan.[90] The Poultry Improvement Program solved the disease problem by supervising hatcheries, testing for pullorum disease, removing diseased eggs, and grading chicks.[91]

ᏇᎲ

Over the course of the 1930s, cotton farmers in Georgia's Upcountry began trading cotton production for poultry raising, a process facilitated and promoted by the federal government and national feed companies. With the backing of the companies, proprietors of country stores advanced chicks and feed to farmers, just as they had advanced seed and fertilizer to those raising cotton. As poultry farming grew during World War II, poultry became the chief source of income for many farm families. As Upcountry farmers traded cotton for poultry, they would soon find that this shift neither amounted to diversified farming nor marked an end to the problems that accompanied cotton production and the crop lien. Indeed, poultry raising began to reinforce and even heighten the problems of a one-crop economy. As Upcountry Georgians increased their poultry production, they could not foresee the shape and degree of dependence and indebtedness that lay ahead. Even so, enthusiasm led the day.

On the eve of World War II, a local newspaper triumphantly announced, "The fuzzy down of the baby chick has all but ousted the fleecy lock of the cotton boll from its pedestal as chief money crop of Hall County." This zeal was justified at the end of the war. Before the war, poultry products in Hall County, the center of Georgia's poultry industry, were valued at roughly $120,000, which represented about 14 percent of all the farm products sold from the county that year. Ten years later, poultry products in the same county were valued at almost $6 million, representing 86 percent of all the farm products sold from the county that year.[92]

In 1939, farmers in Georgia produced 1.6 million chickens; but by war's end they produced 30 million chickens a year and 47 million in 1949.[93] "The largest source of farm income," the county agent of Hall County wrote in 1946, "is derived from the production of broilers."[94] In 1947, chicken raising was the chief source of income in ten counties in the Upcountry, and year after year farmers increased their production.[95]

By 1950, chicken had become the Upcountry's principal cash crop.[96] Whereas in 1935 Georgia farmers had produced merely five hundred thousand chickens, fifteen years later that number had grown to sixty-three million, making poultry one of the state's most important farm products and Georgia the nation's second largest chicken-producing state. Even so, the poultry industry neither gave succor to all Upcountry Georgians nor preserved the farm.

"The collapse of the regnant cotton culture of the old Southeast" was at hand, declared the reformer Will W. Alexander in 1936. "Cotton farming in this area," he wrote soberly, "is doomed."[97] Indeed, cotton farming in the Upcountry would eventually meet its demise, but its death would be slow. It was World War II, not the New Deal, that halted cotton production in the Upcountry. Cotton export opportunities diminished with the war, and the government called on farmers to raise less cotton and more food.[98] In 1942, county agents in the Upcountry pressed farmers to reduce cotton by at least 20 percent.[99] In 1944, the county agent for Cherokee County reported that cotton was no longer the largest source of farm income; farmers earned the majority of their keep from chicken.[100] Jackson County's extension agent John L. Anderson reported the same findings in 1945. "The cotton crop this year is the smallest planted in a number of years." "The poultry industry in Jackson County exceeded the cotton crop," Anderson wrote, observing that the industry only continued to expand.[101] In 1946, Anderson reported that cotton acreage harvested had fallen "to the lowest in [the county's] history."[102] At long last, cotton gins began to sit idle. "Of the 23 cotton gins in Hall county only 11 operated last year," Hall County's extension agent L. C. Rew reported in 1946. He predicted that even more gins would fall into disuse in the next year.[103]

In 1929, Rupert Vance surveyed the southern landscape and saw no end in sight to the cotton production that had impoverished the South. For some time, he waged an unsuccessful war against cotton production, the crop lien, and its close relatives sharecropping and tenancy, which he considered to be the root of southern poverty.[104] Far from dismantling the crop lien system that had forced so many farmers into poverty, poultry production was built as an adaptation of that very system. In a word, merchants applied the crop lien to poultry production.

In 1927, Georgia State College of Agriculture professor J. H. Wood prophesied that poultry was "destined to become one of the state's main industries at an early date." After all, he wrote, poultry "offers the most desirable avenues for investment and hence provides an opportunity for diversification which should not be overlooked or neglected any longer by our people."[105] Poultry raising never became the route to diversification that Wood envisioned. In the decades to come, Upcountry Georgians became dependent on chicken in the way that they and their predecessors had depended on cotton—a dependence that—for many—led to poverty and indebtedness.

CHAPTER 2

World War II and the Command Economy, 1939–1945

IN 1941, as the United States entered the Second World War, Claude Wickard, the secretary of agriculture, urged farmers to raise food to ensure "a full dinner pail for every American and adequate diets for the people of nations resisting aggression."[1] Farmers in the Upcountry followed his edict and over the course of the war increased poultry production exponentially. The federal government played a crucial role in the birth of the region's commercial poultry industry, and a combination of war food policy and the command economy transformed chicken farming. Over the course of the war, chicken farming underwent a revolution in the Upcountry, and the commercial poultry industry was firmly established and thriving by war's end.

In 1942, the government rationed red meat and consumers began to purchase more poultry. Soon thereafter, the army became the prime buyer of chickens. The government demanded standardization, inspection, and increased production and played an essential role in the growth of the industry by funding extensive research to increase production, decrease growing periods, and fight disease. This research and the resulting technological advances served as a direct subsidy to the poultry industry and by the 1950s had transformed the poultry industry into what some labeled as the most advanced form of farming.[2] The range of new government policies and regulations made growth possible, but that growth came at a cost; small farmers soon found it hard if not impossible to keep pace with expensive technology, government regulations, and standardization. In most cases larger enterprises were able to keep up but smaller enterprises found it difficult to adhere to these regulations and in many cases ceased production. This process, later referred to as a "shakedown," slowly began during the war and set the stage for the revolution that took place in the decade that followed.[3]

Before the war, poultry products in Hall County, the center of Georgia's poultry industry, were valued at roughly $120,000, some 14 percent of all farm products sold from the county that year. Ten years later, poultry products in Hall County were valued at nearly $6 million, representing 86 percent of all the county's farm produce.[4] By 1950, chickens had become the Upcountry's principal cash crop.[5]

Over the course of the war, the poultry industry became the chief source of income for Upcountry farmers. In 1940 Upcountry farmers raised 2.8 million chickens, and two years later 10 million.[6] Upcountry farmers produced 16.5 million chickens valued at just over $9 million in 1943.[7] In Forsyth County in 1940, farmers built roughly one hundred chicken houses and produced approximately $1 million worth of chickens and eggs.[8] One year later, in Forsyth County farmers earned $1.35 million from the sale of chickens.[9] Chicken revenue surpassed cotton revenue in the Upcountry in 1941, a landmark moment for Upcountry farmers. Cherokee County farmers earned over $1 million from chicken production, compared to the $350,000 from the sale of cotton.[10] Over the course of two short years, commercial poultry had become the main source of income in Cherokee, Forsyth, and Hall counties.[11] Georgia ranked seventeenth in the nation in chicken production in 1937, fourth in 1943.[12]

Part 1. Revolutions Down on the Farm

The problems of the first decades of the twentieth century followed Upcountry Georgians into the 1940s. "A malnourished and unhealthy people can never be prosperous," wrote W. V. Chafin, Hall County's agricultural agent, not long after the United States entered the Second World War. In Chafin's estimation, if Upcountry Georgians were to contribute to the war effort, they needed to "become sturdier in body, steadier in nerves, and surer in living."[13] In 1940, tenants and sharecroppers still operated two-thirds of all farms in the Upcountry, and many moved yearly in search of better conditions.[14] New Deal allotments had curbed but not eliminated cotton production. County agents agreed that cotton depleted and eroded the soil and was to blame for persistent poverty.[15] Change was under way.[16] The number of bales of cotton produced in the Upcountry declined during the 1940s, and by 1947 cotton, in the words of one county agent, had become a "minor cash crop."[17]

The coming war and more specifically the draft highlighted the poverty that persisted among rural people. The army deemed 53 percent of draftees in Georgia unfit for military service or fit for only limited service.[18] Medical professionals primarily attributed the poor health of draftees to malnutrition. Georgia boys were not alone. The Selective Service disproportionately rejected farm youth throughout the nation for physical, mental, and educational defects.[19] Many of Georgia's farmers suffered from pellagra and hookworm and had little or no access to medical care.[20] In Georgia, a state nutritionist assumed that health problems extended well beyond men of draft age. "We may take it for granted," the state nutritionist wrote, "that the rest of the population is in no better physical condition than the young men."[21]

The draft and war industries drew farmers off the land, setting in motion a migration that reshaped the South. Thousands of men left the Upcountry for military service and thousands more began working in the defense plants. "The impact of the war has reached into every farm home directly or indirectly," county agent H. A. Maxey observed.[22] High wages drew many farmers into defense plants and defense-related industries,[23] and Upcountry farmers worked in a parachute plant in Gainesville during World War II.[24] Between 1940 and 1944, roughly 1.6 million farm people throughout the nation served in the armed forces, and another 4.6 million discontinued farming and moved to towns and cities.[25] In Georgia, industry and the armed forces not only pulled farmers from the land but also pushed farmers off the land by seizing farmland to build military bases.[26] One home demonstration agent observed that an entire county was "practically all taken over by Fort Benning."[27] In the words of historian Pete Daniel, "the armed forces and defense work became the re-settlement administration for rural southerners."[28] Some county agents tried unsuccessfully to keep farmers and farm laborers on the land but found that many tenants and sharecroppers insisted on seeking out the high wages that industrial work promised.[29]

As farmers headed to war or left the land for industrial work, landed farmers who were left behind complained of a crippling labor shortage. County agents, who had long been the handmaidens of landed farmers, attempted to stem the migration and maintained that the Upcountry suffered a labor shortage, a result of the high industrial wages that drew tenants and share-croppers from the land.[30] Most cotton farmers declared that they simply could not compete with industrial wages. In Jackson County, the cost of farm labor rose by 20 percent during 1944.[31] Many landed farmers complained that a labor shortage hindered crop production. In Georgia the farm labor supply declined by at least 30 percent between 1937 and 1941. Landed farmers seemed equally perturbed by the fact that such labor as they could find was not cheap. One Georgia landowner found it difficult to find tenants to work his land. His children had sought employment in textile mills, and his former tenants had left to work in war industries. With no ready source of cheap labor, he explained that he wanted "a small panic." "No big panic," he cautioned, "but just enough of a panic to get people back on the farm so that we could get tenants." He wanted cotton prices to fall a bit and noted, "if we could get labor for 50 and 75 cents a day we could get along fine with 10 cent cotton." Many Georgians welcomed the high wages and industrial jobs that accompanied the war, but not everyone. The same Georgia landowner longed for Depression-era conditions, when he reaped the benefits of high unemployment. "When the Hoover bad times come I just quit working," he recalled. "I didn't have to work because I could get all the tenants I wanted here."[32]

In the face of high labor costs and a decline in farm laborers, county agents and landowners herded women, children, and the elderly into the fields and sought to restrict the movement of African American workers. Women and children worked long hours, one county agent reported, adding that they displayed "as fine a type of patriotism as ever a group displayed." "Never has a group responded more gallantly to a cause than did" the farm men, women, and children during 1945, according to another county agent.[33] In the Up-country, schools closed during September and October so children could help gather the cotton crop.[34] African American children bore a disproportionate burden; during the 1944 cotton harvest, Jackson County closed six white schools and all of the black schools.[35] Landowners found additional means of keeping African Americans in the field. "Some farm area Negro day laborers are told not to have radios in house," a government report discovered. Landowners feared that workers would hear of laws that forbid low wages and long hours. The government report found that landowners, with the cooperation of county officials, tried "as much as possible . . . to hold to a minimum the out-county news received in the county."[36] Discriminatory practices and the work of women and children would not sustain cotton production indefinitely.[37]

Despite high farm wages and the exodus of farm labor, Georgia farmers produced a record cotton crop in 1943. One report explained that fewer workers, coupled with the use of machinery, meant that "the output per worker in 1943 [was] probably the largest on record" and that farm income was the highest on record.[38] County agents took an active role in organizing the sharing of farm machinery. "Farmers became machinery minded," Maxey, the county agent for Cherokee County, claimed, and he maintained that all farmers were interested in labor-saving machinery that promised to relieve "much of the drudgery of farm life and increase the efficiency of the farm."[39] The number of tractors across the country increased dramatically over the course of two decades. In 1909, farmers throughout the nation owned a mere 1,000 tractors; by 1940, over 1.7 million tractors plowed the country's fields and diminished the need for hand and animal labor.[40]

Most USDA bureaucrats and county agents hoped that mechanization and the exodus of farmers would end persistent rural poverty. Claude Wickard, secretary of agriculture, believed that the size of the farm population was a "fundamental problem" that mechanization could solve. Wickard welcomed what he called the "tractoring off of sharecroppers in the South," which he believed was "one of the most striking evidences of a country-wide tendency in agriculture to supplant men with machines."[41] Wickard and other agricultural bureaucrats viewed the movement of farmers off the land as a positive development that would raise farm incomes and end rural poverty. "The trend toward fewer farm people, larger farms, and a larger output per farm worker," Kenneth Treanor, a Georgia extension agent, argued, "would tend toward a more prosperous rural economy in Georgia."[42] Wickard's and the USDA's stance on mechanization and the farm exodus suggests that the USDA was finally realizing its long-standing modernist plan of pushing out marginal farmers and achieving full-scale mechanization. The agency was concerned with supporting large farming operations, to the detriment of small farmers and farm laborers.[43]

Few Upcountry farmers shared in the benefits of mechanization. Tractors and later cotton-picking machinery proved unsuitable for the small hilly plots of land in the Upcountry. Mechanical cotton harvesters were being developed on the Hopson Plantation in Mississippi.[44] The mechanical cotton harvester was a dream that began in the 1890s and passed through numerous iterations, all depending on flat land, not hilly Upcountry land, for success.[45] Explaining the relative lack of farm machinery, Spurgeon Welborn, a county agent and farmer, remarked that "the land is not good, not level" and farmers could not utilize the large machinery used in south Georgia and the Mississippi Delta.[46] Welborn explained that north of Macon "there's not much land that's really conducive to big equipment," adding that the combines and cotton harvesters developed during World War II "don't work in this land up here like they do 1,500 miles south of here."[47] Another farmer, Sanford Byers, echoed Welborn's observations. He recalled a trip to south Georgia, where the land was smoother and farmers used large machinery. Byers noted how machinery changed the nature of work after the war: "I seen that they was getting tractors—two and four and eight row cultivators, cotton pickers." He observed that machinery put day laborers and tenants out of work, explaining that mechanical cotton pickers and cultivators "cut the colored people out from picking cotton—out of a job." Upcountry farmers could not reap the benefits of the new farm technology, Welborn noted. "We can't run that kind of machinery up here—it's too hilly."[48] Labor costs and the lack of laborers made cotton farming increasingly untenable in the Upcountry, and in 1943 farmers planted less cotton than they had in years past.[49]

Part 2. War Food Policy and the Command Economy

As cotton production declined, war food policy stimulated an expansion of poultry production. Government meat rations and the growth of military installations in the South increased the demand for poultry. The share-the-meat ration program, started in 1942, restricted "civilian consumption of meat to not more than two and one-half pounds per person per week."[50] The government mandated that civilians curtail their consumption of red meat, and American consumers began buying and eating chicken. Rationing of red meat proved to be a boon for poultry growers and brought about, in the words of the secretary of agriculture, "a ready civilian market for poultry."[51] The war brought factories, bases, and new markets for poultry to the South. Arthur Gannon, the state's chief poultry extension agent, determined that Upcountry farmers could look forward to a bright future.[52] New markets sprouted around defense projects and drove up the prices and demand for poultry.[53]

At the beginning of the war, the government attempted unsuccessfully to stem the growth of chicken production and increase egg production through price supports. With the passage of the Steagall Amendment in 1941, the secretary of agriculture began to support farm prices. The USDA deemed that commercial poultry production had already increased, making it inadvisable to encourage further growth, and accordingly supported prices for eggs and some other farm products, specifically excluding chicken.[54]

The USDA and the War Food Administration reasoned that chicken was unsuited for shipment abroad and argued that eggs were ideal for shipping and for converting feed into "defense foods."[55] The government anticipated a feed shortage and took measures to increase egg production and restrict chicken production. The USDA cautioned that livestock and poultry production had advanced ahead of feed production.[56] County agents warned of an imminent feed shortage that threatened production.[57] Southern poultry farmers told government officials that they were having a difficult time securing adequate high-protein feed.[58] Gannon recommended that farmers fill coops with hens, not chicks.[59] He told farmers that their flocks of laying hens would aid the emergency food situation in England, "where the need for dried eggs was urgent."[60] The government urged farmers to revise and increase egg production goals.[61]

When commercial chicken farming started, farmers used terminology that denoted both seasonality and size. Spring chickens were chickens hatched and raised in the early spring and were therefore a seasonal crop. The terms "fryer," "broiler," and "roaster" referred to the size of the bird and the amount of cooking time needed. By the start of the war, chickens were no longer a seasonal crop, and farmers had begun producing chickens that did not vary greatly in size and cooking time. Most chicken farmers in the Upcountry raised chickens that were about three pounds, larger than fryers but smaller than roasters. The term "broiler" was eventually used for all commercially raised chickens.[62]

The Steagall Amendment used the term "broilers" to designate young chickens that weighed less than three pounds. At the start of the war, the USDA wanted to discourage the growth of broiler farming, not only because the industry had grown rapidly but also because it deemed broilers unsuited for school lunch programs and canning, because of the low yield of meat.[63] To promote production of larger birds, which the USDA desired, the government established ceiling prices and price differentials: roasters weighing more than five and a half pounds were 34.5 cents a pound, broilers 28 cents. The price differentials encouraged widespread violations, and price ceilings led to arbitrary terminology upgrading. Dealers and farmers began simply selling their broilers as roasters, demanding the higher price and taking advantage of the terminology to skirt government regulations.[64]

In 1942, the government reversed its previous policy and established the Grow More Poultry Program. It sought to supplement civilian meat supplies with broilers, setting a goal of two hundred million three-pound chickens for the winter of 1942.[65] The same year it set a price ceiling on broilers to manage growing consumer demand and prevent price gouging.[66] In the Upcountry, farmers responded angrily; five hundred of them attended a meeting and "expressed themselves in no uncertain terms on price fixing and cost of production," arguing that the ceiling price fell below the cost of production."[67] Gannon reported that the predicted food shortage, the limited amount of red meat, and government calls for increased broiler production resulted in "a big increase in the interest in poultry raising."[68] In the first months of 1943, consumer demand for broilers exceeded the supply.[69]

With the passage of War Food Order 119, in December 1944, the USDA and the War Food Administration shifted from encouraging chicken production for civilian consumption to procuring chicken for the military. The military began purchasing large quantities of poultry for use on domestic bases and abroad and required poultry farmers to sell to government-authorized poultry processing plants.[70] As the government became the prime buyer of poultry, it mandated that farmers and processors expand and standardize their production by mandating everything from the minimum weight of chicken to the required size and shape of their sheds.[71]

The USDA set prices for broilers but not for feed and chicks, which encouraged the emergence of a black market. Farmers in the Upcountry were angered by government policy that increased production costs but not returns. County agents discovered "widespread disrespect for the law."[72] Farmers began to sell broilers on the black market, and the military reported that it was having a difficult time procuring sufficient quantities of meat.[73]

Farmers' illegal chicken sales led to a climate of lawlessness in the Upcountry. The USDA patrolled the roadways in search of broilers bound for the black market.[74] War Food Order 142, effective August 1945, limited the transport of broilers to a distance of one hundred miles. The order was issued in an attempt to stop truckers who were illegally carrying live broilers from farms to urban markets.[75] However, truckers soon realized that the War Food Administration did not have the authority to stop and search trucks. Officials reported that black market chicken truckers refused to stop for inspection, insulted agents, and threatened them with violence.[76] Many claimed that truckers used old moonshine routes to haul broilers to the black market. In urban centers, black market poultry sold for about one dollar a pound.[77] One historian explains that government regulations and the rise of the black market enabled Upcountry farmers to pull ahead of poultry farmers in Delmarva, the peninsula where Maryland, Virginia, and Delaware meet, which on the eve of the war was one of the largest chicken raising areas in the country. Since trucks in that area had very few routes to enter and exit the peninsula, the government could easily control the movement of broilers. The hills and backroads in the Upcountry, by contrast, suited the black market truckers. Georgia farmers and dealers made important contacts with urban distributors as Delmarva farmers lost their lucrative contacts.[78]

Part 3. Regulation and Research

Government demand stimulated the expansion of hatcheries, commercial facilities where eggs hatch under artificial conditions. Hatcheries were crucial to industry growth because they ensured a steady supply of healthy chicks.[79] In 1940, there were 138 hatcheries in the state, mainly in the Upcountry, and by 1946 they numbered 187, with expanded capacity. Georgia hatcheries produced 1.7 million chickens in 1932, 3.5 million in 1940, and 9.5 million by 1946.[80] As capacity grew and confinement encouraged diseases, the USDA began to implement breeding and disease control programs.[81]

Government regulation also stimulated revolutionary growth of processing plants. As of January 1945, the army had become the prime buyer of broilers and purchased only processed poultry, halting the sale of live birds. "The territory was largely built on a live [poultry] market," one county agent wrote, and agents were charged with explaining to farmers the need for dressing and freezing poultry for export.[82] In turn, the number of processing plants grew. Upcountry Georgia's first processing plant opened in 1941.[83] Demand and later government regulation attracted the attention and capital of national processing corporations, which came to the region to build packing plants.[84]

A year after the first processing plant opened, Wilson and Company of Chicago moved into the Upcountry.[85] Hall County's five poultry dressing plants processed thirty thousand birds a day in 1944, reaching sixty thousand one year later. In 1946, Hall County had six dressing plants that processed seventy-five thousand birds per day.[86] These plants had an enormous impact on area employment, and both chicken houses and processing plants created enduring environmental issues.

As with other elements of the poultry business, the military set the standards for processing broilers. Prior to military procurement, processing plants prepared chickens in three ways: New York, dressed and drawn, and cut for table. The New York process, so named because markets in New York City generally sold this type of processed chicken, was simply the removal of blood and feathers and was common before World War II. With the dressed and drawn broilers, processors followed the New York process and then removed the entrails but left the poultry whole. The cut for table process incorporated all the aforementioned processes, and processors disjointed the broilers as well. The military required processors to station federal inspectors in the plants and demanded broilers cut for table, which then became standard.[87]

Military regulations increased costs for processors, forced out small operators, and stimulated consolidation.[88] In 1943, local dressing plants appealed to the Extension Service, asking for more time to meet government requirements. The "130 degree slight scald," a technique used in poultry processing, increased the cost of production by 50 percent.[89] These wartime measures continued during the postwar period. After 1945, only government-authorized processors could sell poultry.[90] Small processors had a difficult time earning classification, and some complained of corruption. One small processor wrote to the USDA and argued that the regulations gave large processors an unfair advantage. The War Food Administration's regulations, he wrote, "have been arranged to help big business and discourage little business." The processor maintained that heads of large firms had special connections to officials at the War Food Administration and the Office of Price Administration and used them to secure army contracts.[91] "I could prove to you," the small processor wrote to the USDA, "that the large dealers on advisory committies [sic] . . . have formed a collusion to wipe out competition."[92] It was common knowledge in the Upcountry that Extension Service officials and Jesse D. Jewell, one of the largest processors, made no effort to hide their close relationship. Officials from the service gave Jewell extensive support and frequently touted his innovations.

As part of its drive to standardize poultry production, the USDA funded additional research on freezing, processing, feed, poultry disease, and packaging—all of which benefited Jewell and other large producers. In 1943, for instance, the government investigated freezing facilities in the Upcountry, determining that one car of dressed poultry left Gainesville each day. L. C. Rew, the extension agent there, maintained that poultry production outran the capacity of freezing facilities.[93] In 1943, agricultural agents worked with Jewell's company to develop new packaging and freezing techniques.[94]

Government-funded research continued after the war. The federal government built a laboratory to study poultry diseases in the Upcountry in 1946.[95] The city of Gainesville collaborated with Hall County to build a laboratory to conduct experiments on poultry disease. The state of Georgia kicked in $70,000 a year for the lab's operations. A news article about the new lab quoted county officials who touted the lab's research as important for decreasing the loss of chicks. In addition to saving chicks, they argued, experiments conducted by animal pathologists there would improve public health by fighting diseases spread by chickens, including typhus fever and tuberculosis. The article hailed the lab's potential for "combating such menaces to community welfare." According to the article, poultry production had been a boon. Farm income had exceeded $6 million in 1945, compared to $1 million in 1940. The command economy transformed the lives of many white farmers in the Upcountry. Five hundred new farm homes dotted the countryside and more than thirteen hundred farms were now electrified.[96]

In the Upcountry, African American farmers were not beneficiaries of the command economy. Likewise, they did not reap the rewards of government research. The USDA's racist policies and discriminatory practices pushed African American farmers to the sidelines. The agency systematically discriminated against African American farmers, who did not share in the new poultry wealth as they received little aid from government agents, who focused exclusively on white farmers. R. J. Richards, the Georgia poultry marketing specialist, explained that during 1941 the agency's work among African American farmers "had little or no direct dollar and cents value to those concerned." He went on to explain, "Practically no volume of eggs was handled by these club members as the chickens kept by the negroes were very inferior and small in number."[97] In her 1941 annual report, Camilla Weems, the sole Negro Home Demonstration Agent for the state of Georgia, explained that African American farmers in Georgia lacked "poultry equipment suitable for raising poultry commercially." They needed "proper poultry houses, poultry feed and good breeds of poultry, and the proper knowledge of how to make poultry raising commercially profitable."[98] Added to these disadvantages, some 95 percent of black farmers were tenants or sharecroppers.[99] They continued to raise yard chickens but seldom entered commercial scale.[100] African American farmers watched the Upcountry's poultry boom, but with little access to credit and information, they were not beneficiaries of the command economy. It is no wonder that Gannon reported in 1948, "No negroes have hatcheries and few have commercial flocks in the state." For more than a decade, Gannon's extension agents had systematically discriminated against African American farmers, effectively barring them from commercial poultry production.[101]

Part 4. At War's End

A year before the end of World War II, in August 1944, Secretary Wickard spoke before the House Special Committee on Post-War Economic Policy and Planning: "I want to sound a warning against any belief that there can be any sizeable back-to-the-land movement after this war." After the war, he explained, "agriculture will offer no large scale possibilities. . . . We cannot afford again to think of agriculture as a refuge or national poorhouse." Acknowledging that many enlisted men came from farms, he wanted to bar veterans from returning to the countryside and seized on the farm exodus that had taken place during the war as a means of ending rural poverty in the American South.[102] From the start of the twentieth century, the USDA had unsuccessfully attempted to end rural poverty in the South. The agency used World War II as an opportunity to end rural poverty, not by solving the persistent problems that plagued southern farmers—debt, the crop lien, malnutrition, poor sanitation, soil erosion and infertility, glutted markets, overproduction—but by encouraging the exodus of farmers and pursuing policies to prevent the return of veterans and war workers to the land in the postwar period—a rural slum clearance of sorts. In the aftermath of World War II, the USDA's desire to prevent a "back-to-the-land" movement coincided with the labor demands of the new and growing military-industrial complex. During World War II, the military-industrial complex reshaped farming, the agricultural economy, and social relations in the American South, and the USDA joined to clear the land of sharecroppers and small farmers and retune the land to science and technology.

The defense industry and the military did not merely pull farmers from the land; the government also pushed farmers off the land by seizing lands to build military bases and defense plants.[103] In 1943, the Bureau of Agricultural Economics (BAE, a division of the USDA) reported that the armed services had purchased or leased roughly twenty million acres of land throughout the United States, including six million acres of farmland; fifty thousand farm families had been displaced. During the New Deal, projects such as the Tennessee Valley Authority also displaced thousands of farmers. As part of postwar planning, the BAE was charged with reviewing if and how land would be returned to farmers. In preliminary planning, the agency determined that only three million of the six million acres of farmland seized by war industries and the military would be returned. One report explained, "numerous sites were submarginal for farming and should be devoted to non-farm uses."[104]

Wickard and other agricultural bureaucrats celebrated and argued that it would raise farm incomes and end rural poverty. The USDA seemed convinced that it was finally doing away with—in Wickard's words—"little farms for little people,"[105] "subeconomic" farmers,[106] and "those who saw agriculture as a way of life."[107] The USDA's stance on mechanization and the farm exodus suggests that it was well on its way to becoming concerned with supporting large farming operations, to the detriment of small farmers and farm laborers.[108]

USDA policies and the labor demands of the postwar military-industrial complex threatened planter hegemony. In the aftermath of the war, tenancy and sharecropping and the social relations and economic system that had existed in the South since the 1870s began to crumble. To be sure, the landed elite resisted the revolution.[109] But in many ways the war created a new world and new opportunities. In Georgia the farm labor supply began to decline. County elites had gained enormous power by sitting on USDA committees that assigned acreage allotments, made loans, and distributed payments. Large farmers still guided the process of change.

As they began to plan for the postwar period, USDA bureaucrats wrote in somewhat alarming terms about the need to prevent migration back to the farm. A 1943 report read, "At the close of the war there will be millions of workers who will adjust their employment, involving relocation in a large number of cases. Many will return to the area from which they came. If there is no plan to expand the industries in the areas of underemployment these workers will move into farming and further complicate the farm problem."[110] But USDA bureaucrats had only vague hopes and vaguer ideas about where returning veterans and war workers might find work. And while the department was convinced that it wanted to reduce the number of farms and increase the size of those farms remaining, it could not provide the millions of jobs that war workers and veterans would soon need. However, in the aftermath of World War II, the USDA's desire to prevent a back-to-the-land movement coincided with the labor demands brought on by the Cold War and the new and growing defense industries and military bases, which located themselves in the South and funneled billions of dollars into the region.[111] In addition, many rural people who moved to cities during the war for work had no intention of returning to the countryside. Having seen cities, they were hesitant to return to an isolated Upcountry farm with no running water or indoor plumbing.

Many postwar rural southerners, then, did not return to farms but worked in military installations and defense plants. The Cold War hastened and strengthened the South's dependence on the military-industrial complex, and the region drew massive amounts of federal defense spending.[112] As federal dollars poured in, southern towns became cities overnight and fields became military bases and nuclear testing sites: Oak Ridge, Tennessee; Savannah River Site; Mobile, Alabama; New Orleans.[113] Historian Bruce Schulman writes that the South "remained the nation's boot camp throughout the postwar era."[114] As the population of cities grew the rural population declined.[115]

In January 1946, the army abruptly cancelled all orders for poultry, leading to the largest poultry stores on record.[116] At the same moment it ceased buying poultry, the government also lifted meat rations, and consumers began purchasing red meat again, leading to a drop in poultry sales.[117] In 1946, there were over 353 million pounds of frozen poultry in government cold storage holdings, and the price of broilers dropped precipitously.[118]

Nonetheless, Upcountry farmers expanded broiler production.[119] In the immediate aftermath of the war, veterans began to request advice from the USDA about how to enter broiler raising. Officials at the USDA cautioned against entering the industry, but their recommendations largely fell on deaf ears. "I read in the papers that you had 1000 baby chicks sent to you & that you didn't want them," one veteran wrote to the USDA. "I am to be discharged in a few days," the veteran explained, and he intended to go into chicken raising. "I would be very happy to take these 1000 chicks off your hands, if they are still alive. . . . I will pay the express charges when they arrive."[120] Another veteran wrote to the USDA asking if a small producer could make a profit, and the official who responded cautioned against entering the chicken business: "It would seem a mistake to start such an enterprise, particularly on an extensive scale."[121] Another official responded to a query explaining that the USDA could not "give its endorsement to [a] proposal of encouraging disabled veterans to enter the poultry business at this time."[122]

※

When veterans returned to the South after the war, they saw radical changes in the landscape and economy. Tractors filled the fields, and military factories and military bases could be found in every state and on the outskirts of most southern cities. Veterans who returned to the Upcountry may not have encountered a world that seemed drastically different from the one they had left behind after the bombing of Pearl Harbor. The Upcountry's small farmers had not been "tractored off" their land, and they were now producing broilers, a crop that seemed to have brought wealth to a once impoverished region. It probably seemed as though the Upcountry farmers had weathered all the drastic changes that had transformed the South. There were still small farmers who owned their land and worked for themselves. Broiler production seemed to have staved off some of the revolutionary changes occurring throughout the South and seemed to insulate Upcountry farmers from change and industrialization. Veterans could not see that farming was slowly but surely becoming capital-intensive and that the cost of production for broilers was only rising.

The growth of the industry during the war brought both optimism and unexpected profits to farmers. At the start of the war, Sanford Byers, an Upcountry farmer, began building chicken houses when his cotton crop had been laid by. Byers recalled the first time his feed dealer came by to gather the first batch of chickens. The feed dealer drove a Ford truck, a symbol of wealth and a sign of good things to come. Byers's feed dealer returned from Atlanta and wrote him a check for $165. "That was the most money I'd ever seen in my life." Byers quickly reinvested the money, building another chicken house and buying a Ford truck of his own to haul birds and feed.[123]

By the end of the war, many Upcountry Georgians had traded cotton farming for poultry growing. Farmers cut cotton acreage, expanded chicken houses, built new houses, and began investing in feeding, watering, and brooding devices. The change in crop did not guarantee an end to the problems that accompanied cotton production and the one-crop economy. Farmers could not know what lay ahead. In time, broiler raising would reinforce and even heighten the problems of a one-crop economy. Upcountry Georgians increased their poultry production but could not imagine the shape and degree of dependence and indebtedness that lay ahead.[124]

The poultry industry seemed to bring wealth to the region, but county agents warned farmers against becoming too reliant on poultry as their sole source of income and reported that farmers were taking on more debt than in the past. Immediately after the war, H. A. Maxey, Cherokee County's extension agent, wrote, "One of our greatest problems is to keep people farming and to keep them from going too deep into the poultry business."[125] Upcountry poultry farmers began to build new chicken houses and installed electric lights and running water in the chicken houses to save time and labor.[126] In the postwar years, they had access to a whole new range of machinery and equipment. As farmers mechanized, they incurred debt.[127] Over time the treadmill of more chicken houses, expensive feed, updated technology, and crushing debt ruined many poultry producers.

To be sure, broiler production seemed attractive to veterans because of the prosperity they observed when they returned from war. Despite the fall in demand, farmers continued to raise poultry rather than reverting to cotton production. "The largest source of farm income," the county agent of Hall County wrote in 1946, "is derived from the production of broilers."[128] In 1947, broiler raising was the chief source of income in ten counties in the Upcountry, and year after year farmers took steps to increase production.[129] County agents reported an unregulated boom in production and observed that farmers began to improve their standard of living.[130] Maxey maintained that farmers in Cherokee County were now a "better satisfied rural people." In 1949, 92 percent of farm homes in Cherokee County had electricity, although only about 25 percent of homes had hot running water. Those farmers who finally had electricity began to purchase all of the goods that one associated with middle-class American homes, including washing machines, radios, refrigerators, and electric irons; they even covered their kitchen floors with linoleum.[131]

Over time it would become clear that a tough bargain had been reached. Upcountry poultry farmers stayed on their land and to a degree remained independent. Unlike the pattern of development found on southern neo-plantations and in midwestern and western agribusiness, heads of industry in the Upcountry neither consolidated land nor forced Georgia farmers into wage work.[132] Poultry firms did not strip farmers of their means of production. Indeed, they took the reverse course and demanded that farmers invest in poultry houses and poultry feeding machinery. As contract farming became the norm, poultry firms set the precedent that poultry growers would assume much of the capital investments for the expansion of the industry. Ironically, poultry growers' dependence soon became rooted in the very fact that they owned the means of production.

CHAPTER 3

Taking Over: Integrators and the Birth of the Modern Broiler Industry

IN THE YEARS following World War II, poultry integrators developed an ingenious and highly profitable business model that consisted of three components: quasi-vertical integration, the feed conversion contract, and the forced sale of broiler housing and machinery to contract farmers. This business model served as the foundation of the modern poultry industry. Not surprisingly, integrators declared that this new business model protected farmers from risk and promised that it would bring wealth to all. It did not. In 1951, Georgia became the leading poultry producing state in the nation. This triumph came at a very high cost—paid by farmers. By 1967, broiler farmers were providing more than 60 percent of the capital to run the million-dollar poultry industry, had no say in production and marketing decisions, and earned so little that they were forced to work off the farm to make ends meet and feed their families.[1]

In the space of a decade, integrators, so named because they had vertically integrated the industry, revolutionized the poultry industry, introducing a business model whereby poultry farmers produced broilers under contract and were paid based on how efficiently they raised broilers. Key to this practice was the fact that the industry was not in fact fully integrated. Poultry integrators did not actually own the machinery and housing in which their broilers were raised. Instead, integrators mandated that farmers, some of the lowest paid workers in this production chain, purchase state-of-the-art machinery and broiler housing. Moreover, integrators sold the chicken houses and machinery to farmers and financed those very loans. Integrators increasingly demanded that farmers repeatedly install the newest equipment and construct the newest and largest housing.

The creation of the vertically integrated poultry firm and the implementation of the feed conversion contract did not take place at a fixed moment. Therefore, it is difficult, if not impossible, to identify a clear chronology of the development of the new business model. We can neither point to a moment when farmers and integrators sat down and agreed on the feed conversion contract nor identify when the large firms made the decision to pursue quasi-vertical integration, and subsequently pushed much of the cost of raising broilers onto farmers. We do know that the decade following World War II marked a period of swift standardization within the industry. In the process, the big guys rose to the top and the little guys exited the game. Farmers either mechanized and expanded production capacity or refused to "innovate" and got locked out of the business altogether. Vertical integration and the feed conversion contract came into being through thousands, if not hundreds of thousands, of small and large negotiations and compromises between farmers and integrators.

There were clear winners and losers. If we use profits as the barometer of success, then it is clear that the new business model disproportionately bene-fited integrators, at the expense of farmers. Perhaps the most difficult piece of the story to understand is how and why farmers agreed to contract terms that were so clearly and astoundingly unfair and unreasonable and allowed inte-grators to grasp the upper hand after World War II. Integrators set the terms of the contract, demanding that farmers invest in housing and machinery, and simply refused to contract with any farmer who declined their terms. So farm-ers were left with two choices: acquiesce to integrator demands and continue to raise broilers or reject the terms of the contract and cease farming. In other words, they could choose to eat or to starve. Integrators benefited from the fact that they were building their business in a poor region where farmers simply did not have the means to make alternate economic choices.

Jesse Dixon Jewell cast himself as the founding father of the broiler indus-try. In interviews, he explained time and again that he had turned cotton farm-ers into broiler growers, and in so doing saved farms and farmers. According to his narrative, he built an industry essential to the economic well-being of the region and its farms. By 1952, Jewell contracted with roughly 600 broiler farmers, and his business employed upward of 550 wageworkers. This was no small feat, and the size and reach of his business help to explain why he was viewed as the industry's founding father. Interviewed by a journalist for the *Los Angeles Times* in 1952, Jewell touted that one of the great things about his business was that "so many other persons can share the wealth."[2] However, this characterization was far from the truth. Jewell had perfected business practices that made poultry central to the economy of the Upcountry but, in the pro-cess, impoverished farmers.

Part 1. Quasi-Vertical Integration

After the war, Jewell and other feed dealers set out to build quasi-vertically integrated poultry firms, where they owned all components of the business except the houses in which farmers raised broilers. It is important to label these enterprises as "quasi" or "semi" vertically integrated enterprises because integrators' choice not to purchase broiler houses was very intentional, and in the end very lucrative. In 1957, Jewell made it very clear that he himself did not and would not raise broilers, nor would he own broiler houses. This part of the business was, in his words, "getting more unprofitable." That was an understatement. Farmers took on unprecedented debt to build single-use broiler houses. Raising broilers was by far the most risky and volatile part of the production chain. Disease could and did kill off entire flocks. Farmers could and did lose flocks of twenty to fifty thousand birds and, under the feed conversion contract, earn absolutely nothing. Knowing the risks of raising broilers, Jewell instead invested in the most stable and lucrative parts of the production chain. He bought and built breeding facilities, hatcheries, feed mills, dressing plants, factories for freezing and preparing processed foods, and a by-product plant, which processed feathers and cartilage to be used in the making of chicken feed. But he did not invest in or own any broiler houses, the costly facilities in which farmers were soon raising record numbers of birds. Instead, he invested his capital in other more promising facets of the business—packaging, precooked foods, and the sale of broiler machinery and housing—and passed the cost of the most risky aspect of production to the farmers.[3]

To create their quasi-vertically integrated firms, soon-to-be integrators began buying up hatcheries at war's end. Many of the smaller hatcheries, which had been established during the war to meet growing demand, discontinued operations or were bought out by poultry integrators. In 1950, poultry extension agent Arthur Gannon observed, "The trend has been toward fewer hatcheries and larger hatcheries." By 1950, Upcountry hatcheries had their own supply flocks and breeding facilities, which meant that eggs were no longer shipped in from out of state.[4] Gannon wrote, "The whole nature of the hatchery business has changed," meaning that integrators now owned and managed the hatcheries, supplying the farmers with chicks.[5]

As of 1949, integrators began to build feed mills and by-product plants.[6] Gannon watched as integrators installed some of the most technologically advanced equipment and began to use feed mixers, protein supplements, and bulk delivery feed equipment. Bulk feed equipment was essential, as it allowed farmers to not lose feed by transporting bags of feed into their silos.[7]

Though time and again, integrators would claim that the free market led to their success, federal subsidies in multiple forms were absolutely key to the grow of the industry. The federal government researched disease preventives and vaccines, and disseminated related information. With excitement, USDA researchers reported that, thanks to their intervention, "broiler producers in

Georgia have available the most technologically advanced and efficient feeds sold in the nation."[8]

Integrators needed trucks to distribute chicks, feed, broilers, and finally processed chickens and prepackaged products. They also needed consistent and reliable transportation. Integrators purchased fleets of trucks, pushing out independent commercial truckers.[9] Incidentally, in the early 1930s in Arkansas, John Tyson entered the poultry industry via trucking, transporting spring chickens and truck crops northward. Tyson remained, according to one historian, "in the shadows as other companies dominated the business." It was not until the 1960s that Tyson Foods became a key player.[10]

From 1945 onward, the number of processing plants in the Upcountry ballooned. As of 1932 there was only one processing plant in the region, but wartime production had driven the growth of plants, and from 1941 to 1955 roughly three plants opened yearly. By 1947, Hall County alone had five plants, each with a daily capacity of over fifty thousand broilers. That same year, one journalist estimated that growers produced a million pounds of chicken meat per week. As of 1955, twenty-eight commercial plants—those processing a thousand broilers or more per day—were in business in the Upcountry. Thirteen of those plants processed nearly three thousand birds per hour.[11] Processing plants then passed along any remaining chicken parts to rendering plants, which ground feathers, cartilage, and skin and supplied feed mills with this product. Feed mills, producing broiler feed, added these renderings to broiler feed. So, broilers ate the remnants of broilers.

Early on, Jewell recognized that profits lay in processed food. Integrators, again with the assistance of the federal government, began to develop value-added processed foods. They reported a growing demand for "cut-up, fresh, tray-packed broilers."[12] Jewell owned processing, freezing, and by-product plants, and oversaw the marketing and distribution of his products.[13] His company developed a range of frozen and prepared products that included, as one county agent explained, "a beautiful package of cut up chicken ready for the pan with recipes for preparation."[14] "Fancy packaging" methods helped to promote sales. Jewell pioneered all sorts of innovations, packing poultry in cellophane and aluminum. He hired marketing specialists to study how best to prepare an attractive and convenient product.[15] In 1957, Jewell explained the innovations he was implementing and how those innovations expanded the market for chicken. "We are getting into the precooked department. . . . We are getting to where we are selling precooked chickens to restaurants and hotels and drive-ins and places where they had never sold chicken before."[16] The USDA subsidized and aided Jewell's innovations. A county agent reported that state and county extension servicemen "worked with the J. D. Jewell Company in experiments on packaging and freezing dressed poultry." A journalist reported that "new methods in fancy packaging for poultry for the further promotion of sales from the Gainesville production area have been studied in Gainesville this week by local poultrymen and specialists in that field from the Experiment station." Thanks to improvements in packaging, one journalist remarked, "the Gainesville area is expecting to lead in offering attractive and convenient handling devices."[17]

Part 2. The Feed Conversion Contract

Along with vertical integration, the feed conversion contract was equally crucial to the founding of the modern poultry industry. Integrators used the feed conversion contract to systematically and incrementally seize control from farmers over all production and marketing decisions. With the feed conversion contract, integrators made two key changes that reverberated throughout the region's poultry industry and soon beyond. First, via this contract, integrators took hold of all marketing decisions, permanently taking any and all marketing decisions out of the hands of farmers.[18] Second, integrators also assumed control of all production decisions. Integrators paid farmers based on how efficiently they turned feed into meat, a metric entirely unrelated to the market price of chicken.[19]

In the postwar period, integrators and farmers used a range of different contract arrangements before arriving at what eventually became the feed conversion contract. Under this system, growers were paid based on how efficiently they converted feed into meat. In 1956, the USDA defined the feed conversion contract in these terms: "A feed-conversion plan in which payment to a grower is adjusted according to an agreed scale that rewards the grower for great efficiency in converting feed to broilers [and] is also close to a wage contract."[20] Yet the system was in fact more complex; this was the mechanism that severed farmers from all marketing and production decisions. The feed conversion contract allowed integrators to build what one study labeled "single decision making unit[s]," wherein the integrators and they alone determined the pace and scale of production.[21] Integrators dispensed entirely with the notion that farmers ought to have a say in growing and selling their crops. Indeed, integrators viewed farmers as impediments who could potentially disrupt the production chain by stalling, deciding to refuse a flock, or slowing production.

Under the feed conversion contract, integrators retained title to the birds, as they had in the decades before, but they ceased selling feed to farmers. Up to this point, feed had been the chief source of revenue for furnishing merchants.

Indeed, it had driven and created poultry farming. Under the new system, integrators paid farmers a fixed rate, based on how much chicken meat a farmer produced per pound of feed.[22] Though drastic changes lay ahead, in the 1950s remnants of the old system remained. All integrators soon required written contracts, but some did not consider a deal finalized without a handshake and a verbal agreement. With or without a handshake, integrated poultry firms, like Jewell's, came to dominate, and the feed conversion contract was firmly established in the Upcountry as of 1953.[23] In the mid-1950s, almost 100 percent of Jewell's growers worked under the feed conversion contract.[24]

Notably the feed conversion rate soon became almost entirely unrelated to the market price of chicken. Prior to the implementation of the contract, farmers' profits were tied to the market price of chicken, and in some cases, even into the early 1950s, farmers retained a minimal degree of control over when and if they would raise a flock. When the market price dropped, they could in some cases opt to slow or halt production, refusing to purchase feed and avoid going into debt.[25] One study reported that each farmer had a "large capital investment tied up in his operation" and therefore "recognized that his livelihood depended upon a fair market return and adjusted his input and output in a responsive manner." A grower, the same study reported, "could readily and in short duration adjust his operation to market realities without suffering significant financial harm." It is fair to say that while this was still the case for some farmers, the report overstated the economic control and independence that farmers possessed. Leaving a house empty meant losing money. That same report explained that the grower's "ability to control input and output meant that breeding farms, hatcheries, processors, and even feed producers had to adjust their operations to some degree to growers' prerogatives."[26] This meant that farmers could technically disrupt production, leaving hatcheries with growing chicks, trucks idle, and processing plants at a standstill. With the implementation of feed conversion contracts, integrators solved this problem; the "single decision-making unit" pushed farmers to the sidelines, severing their relationship to the market and ensuring that the production chain could not be disrupted.

Prior to the early 1950s, there were many variations of what eventually became the feed conversion contract. One study found over twenty variations in use in the Upcountry in 1950 among farmers who grew more than twenty thousand birds. Feed dealers, who became the integrators, tinkered with the contract before arriving at the feed conversion contract. Again, it is worth emphasizing that the integrator incrementally seized control from the farmer over all production decisions. Three precursors to the feed conversion contract— "open account," "no-loss," and "mostly quotation"—help us to understand the evolution that led to the final model. These different incarnations of the feed conversion contract help us to see how integrators gradually managed to seize power over marketing, production, and decision making.

At the end of the war, the majority of farmers and feed dealers continued to use the open account contract, virtually the same model developed in the 1930s, an adaptation of the crop lien system used to finance cotton. Feed remained the main source of profit. The "open account contract" and variations thereof facilitated the massive chicken production that had taken place during the war. Under open account, the feed dealer retained ownership of the birds, as had been the case as of the late 1930s.[27] Farmers and feed dealers shared in both loss and profit. Production decisions were connected to the market price of chicken. Feed dealers' chief source of income was the feed sold to farmers and the interest charged on that feed. Both parties were interested in what the market could bear. As had been the case before the war, most contracts stipulated that dealers owned the birds, while farmers purchased on credit the feed and other supplies necessary to produce the birds. Feed dealers marketed the birds, and if the price of chicken meat fell below the cost of production, the farmer was left in the hole, indebted to the feed dealer. A study of the Upcountry broiler industry in the first years of the 1950s found that the chief financial interest for feed dealers, those who financed broiler production, remained, as it had been in 1930, the sale of feed.[28] That was about to change.

Arguably the most important intermediary step in the transformation of this process was the no-loss provision contract.[29] Under this contract, integrators continued to own the chicks and extend credit for feed. The no-loss provision meant that even if a farmer's cost of production for a flock fell below his return on that flock, he would not go into debt and would lose only his labor and his payments for fuel and the litter that lined the broiler houses.[30] The integrator would essentially forgive and assume the debt. For example, if farmer Smith contracted with Jewell, borrowing one hundred dollars (to simplify, one could include the interest charged by Jewell in this sum). The flock Smith produced sold for eighty dollars, twenty dollars below his cost of production. Under the no-loss contract, Jewell absorbed that loss. Smith was not left in the hole for the cost of feed. He had lost his labor and the cost of fuel and litter. So this system looked acceptable to farmers, whose parents and grandparents farmed cotton and had also remained indebted to furnishing merchants, year after year. Incidentally, the arrangement also looked quite good to some USDA researchers—in their view, paternalistic and benevolent integrators took on more risk to shield the farmers from the whims of the market.

In the early 1950s, USDA researchers conducted a multipart study of the Upcountry's poultry industry. They concluded that with the no-loss provision, the integrator assumed most, if not all, of the financial risk. Under this contract arrangement, integrators did technically insulate their growers from the exigencies of the market. But the no-loss contract came at an increasingly high price to farmers. As they introduced no-loss, integrators increased interest rates on feed and took a larger cut on the sale of the chicken. This made sense to many observers and researchers because, again, the integrator was shielding the farmer from the market, and this protection had to come at a cost.[31] Integrators touted the fact that they were protecting farmers, and USDA researchers took integrators at their word, failing to probe how the no-loss contract might be detrimental to farmers. "The dealer bears the greatest risk when he guarantees the grower a fixed labor income. Under this arrangement, it is natural that risk and credit charges, in the form of margins on feed, chicks, and supplies should be considerable," a USDA report concluded.[32] USDA researchers condoned the business model. According to Jewell, "integrators released and insulated growers from the risk and instability." Growers no longer needed to worry themselves with the market, the price of feed, and so forth. "The only concern in their mind," Jewell made clear, "is how good a job [they] can do, and this is the way they are paid."[33] Perhaps to everyone, including the farmer, the higher markup on supplies was mitigated by the contract's provision that farmers would not lose money if the price of chicken fell below the cost of production. But farmers already detected underlying problems.

As the contract evolved farmers did voice concerns about their increasingly precarious position, but they did not organize to fight the changes. Years after he had left broiler raising, Arthur Flemming reflected on the changes in contracting and the ways in which he surrendered much of his decision making to integrators. "Back when I was in the chicken business on my own," he recalled, "you knowed how many pounds of chickens that you sold the day that they left the farm." Once the feed conversion contract came into to use, Flemming explained that "you don't know till that contractor sends his service man out or mails you that check and mails you the settlement sheet, see." Even though he was no longer under contract, Flemming seemed quite cautious in his remarks, and took great steps to not explicitly criticize the contractor: "I'm not going to say that the contractor cheats you or anything else. But I'm not there to see the chickens weighed. See what I'm talking about? Back when I started off raising chickens, you weighed your chickens at the house. You knew what they weighed, because when he'd weigh those chickens he'd give you the slip of paper and you were standing there watching him weigh them chickens."[34] Flemming suggested that poultry farmers found themselves in a precarious position. Before long, Flemming and other poultry farmers would discover that they faced perilous contract arrangements.

Part 3. "Satellite Activities" and Forced Obsolescence

From the 1950s onward, integrators forced contract farmers to repeatedly purchase new broiler houses and machinery. Integrators developed an entirely new source of revenue; they reaped profits from the sale of housing and machinery and the financing of these sales. This was the third component of the business model that built the modern poultry industry. A 1963 congressional report confirmed that the sale of housing and machinery—"satellite activities"—was central to integrator profits. Integrators, the report explained, "are primarily concerned with maximizing profits of satellite activities."[35] Integrators systematically deemed recently purchased houses and machinery obsolete; one might label their practice as forced obsolescence. Integrators consistently demanded that their contract farmers purchase new, up-to-date, and expensive equipment in order to contract with them, and then returned often less than five years later with more demands for thousands of dollars in "upgrades."[36] Through this "forced obsolescence" integrators secured a captive market for their products.

In a 1956 USDA study, researchers reported that broiler growers "probably adopt improvements more rapidly than almost any other type of farmer."[37] The study was titled, with unintended irony, "Economic Choices in the Broiler Industry." The title's irony lay in that fact that by this point growers had few if any economic choices. They could choose to contract and repeatedly "upgrade" or choose to lose their livelihood. They could shut down operations and lose their homes, land, and farms. Plus, the terms "improvement" and its frequent companion, "upgrade," masked the forced obsolescence. To be sure, the new machinery was technologically advanced. But these terms mask the coercion involved. Farmers were forced to buy, at very high prices, entirely new broiler houses and the latest machinery. Integrators intentionally used the terms "upgrade" or "improvement" because this terminology hid what they were forcing farmers to do. An upgrade or improvement implies that a farmer was making minor changes to broiler houses at minimal cost. This was not the case.

Broiler growers, not integrators, owned the means of production: the broiler housing and machinery. In a strange variation on capitalism, this meant that by the mid-1960s, if not earlier, broiler growers provided more than half of the capital to run a million-dollar industry. Integrators mandated that farmers repeatedly "improve" and "upgrade" their facilities, meaning farmers were required to build entirely new buildings and tear out machinery to replace it with the newest feeding, watering, and heating devices. Integrators contracted solely with farmers who "upgraded" and purchased the latest "improvements." Those farmers who acquiesced to the demands of integrators began to slowly but surely take on wholly unprecedented debt in order to finance these "improvements." Those who did not meet integrator demands lost contracts, were forced out of the business, and left the land.

Integrators boasted that they were providing farmers with the newest technologies to increase efficiency. However, growers, not integrators, covered the bill for that progress. From 1950 onward, broiler growers in Upcountry Georgia paid for the machinery and housing—the very factories—in which they raised chickens. They took on debt to finance this equipment and housing, a practice that became the standard throughout the industry in the South and was later used in agribusiness both nationally and internationally.[38] While they owned their land and the means of production, farmers had no control over production and marketing decisions and remained indebted to the very people who controlled those decisions.

By the mid-1950s, it became clear that integrators focused on "satellite activities," the sale and financing of housing and machinery, not the sale of chicken. At a series of congressional hearings in 1957 investigating the poultry industry, Congressman Charles H. Brown of Missouri interrogated the poultry industry's business model and expressed dismay over the built-in inequities. He observed that integrators' returns grew as growers' returns dropped. Farmers, Congressman Brown stated, were "making less than they were 5 years ago per bird, but they have to furnish even better equipment actually to handle more birds." In his questioning of Jewell, Brown asked, "When a man gets into the market, a man handling 20,000 birds, that man has to have $20,000 to pay for the privilege of getting a job that you say will get him about $1 an hour?" Jewell responded, "Yes." Brown continued, "Just assume that one of the men who is raising chickens for you, in order to do a better job—and this is what I understood you to say, and I want to be corrected if I am wrong—you sell him certain equipment that he needs to do a better job in the production of those birds and you sell at cost; don't you?" Again, Jewell responded, "Yes." The farmer "is investing in that equipment and he is in debt to you for that equipment," Brown stated, again trying to lay bare the precarious position of farmers.[39] The 1963 report on the hearing reinforced Congressman Brown's assessment. Feed conversion contracts, according to the report, "together with an unwise overextension of financial resources, [have] led to a divorcement within the industry of those who desire high production for ancillary profit motives from those who desire normal production for direct and long-term financial well-being."[40] In other words, integrators and growers weren't in it together; the integrators' profits rested on the indebtedness of the growers. Growers were not blind to the fact that they had gotten a raw deal, but now indebted to their integrators, they were left with very few options.

Increased production led to glutted markets and low prices, which plagued the industry in the 1950s (and prompted congressional investigation), but these glutted markets and low prices did not necessarily make much of a difference for Jewell's bottom line. In the Upcountry, growers continued to build structures that held upward of twenty thousand chickens, doubled chicken house capacity, and produced three times as much chicken meat in the 1950s as they had during the 1940s. The reach of the broiler industry in the Upcountry spread as well, as more and more farmers increased production and entered broiler growing. In the early 1950s, poultry prices had fallen to twenty-four cents a pound, and in 1956 prices dropped to as low as fourteen cents per pound. "The broiler industry is now a National problem," lamented H. A. Maxey, an Upcountry Georgia county agent. In 1954, Maxey observed that the situation for farmers was very serious. Farmers in Cherokee County, in his estimation, had "gone through a very hard year with low prices and high costs of production." In 1956, returns to farmers reached record lows, and one county agent observed that "the unbalanced supply and demand for broilers resulted in prices which were disappointing for much of the industry." The government warned of an impending disaster unless integrators cut back on the number of flocks they required. As the 1950s waned, glutted markets persisted, production costs exceeded returns, and bankruptcy among farmers became common.[41]

Integrators did not share the problems of growers. A captive and growing market for machinery and equipment insulated integrators from the increasingly glutted market for chicken that their own actions created. In 1957, Jewell openly admitted that low profit margins for poultry farmers and glutted markets did not necessarily affect his bottom line. Indeed, integrators earned profits irrespective of prices; a firm like Jewell's could lose money on chicken but make up those losses on the sale of machinery. Integration is a "balanced program," Jewell delicately explained. "Sometimes you are able to overcome losses here with profits from this operation there." He ventured that his company had lost money on chickens in 1955 and 1956 but had made profits on the sale of chicken houses and machinery. Integrators like Jewell compelled poultry farmers to subsidize industry growth even as returns to growers fell.[42] If a farmer protested or refused to upgrade, the integrator could simply cease contracting with that farmer and in some cases blacklist him as well, making it impossible for that grower to raise broilers for another integrator. The farmer was then left with costly specialized housing and machinery and massive debt.

ᘓ

Poultry farming had become the basis of the economy in the Upcountry, instituting, once again, a one-crop economy. In year-end report after year-end report, county agents asserted that point. "The business is of such universal interest," the Hall County agent wrote "that we class it as our major project in the agricultural program, since it represents an income of over 90% of our people and economically it means 15 to 20 million dollars this year to our farmers in Hall County."[43] The county agent for nearby Forsyth County reiterated that point. "The broiler industry is the chief source of income for the vast majority of the people in Forsyth County." In 1950, of the county's 11,005 inhabitants, 10,000 were "either directly or indirectly" working for the industry.[44] Another agent reported that poultry was "our main source of farm cash income, involving 1500 growers who produced 26 million birds per year." In that county, related wage work in industries—processing, feed mills, hatcheries, and so on—had "a total pay roll of $1,300,000 per year."[45]

However, Jewell's six hundred broiler growers were not enjoying the wealth. In fact, raising broilers actually threatened the livelihoods of broiler growers because their costs of production were increasingly exceeding their net income. In the early 1950s, it was county agents, not federal researchers, who began to document the chief problem. By that time, growers were providing almost 50 percent of the capital needed to run an integrated firm. Problems for growers only worsened. One county agent reported that "the net income of farmers has not increased as much as gross income figures indicate."[46] As early as 1951 another county agent found that "when all expenses are paid, the net income is generally smaller than they first thought."[47] Fifteen years later, a government study of the nation's poultry industry found that "growers of a firm slaughtering about 20 million birds per year would have invested nearly three times as much as the firm itself ($6.4 million compared to $2.3 million)."[48]

When asked during congressional hearings about these problems, Jewell returned to a hackneyed axiom. Free market capitalism reigned supreme. These men and women, according to Jewell, were free to leave broiler growing at any time. "I consider ourselves a free enterprise," Jewell stated. The growers "are free as birds . . . because they can grow or not grow."[49] Of course, this was not the case. Farmers had assumed thousands of dollars in debt to build structures that served no other purpose than the raising of broilers. Growers and their families were free to work or free to starve.

CHAPTER 4

Broiler Sharecroppers and Hired Hands

IN 1939, Georgia farmer Arthur Flemming entered poultry farming and began to reduce the acreage he planted in cotton. He was one of the thousands of cotton farmers who traded cotton for poultry and helped to build Georgia's poultry industry. Flemming was also one of the thousands of Georgia farmers who followed the edicts of his integrator. He purchased new poultry farming machinery six times in the course of his forty-two-year career. In 1950, Flemming invested $9,000 in a brand-new chicken house that housed nine thousand broilers. It had all of the latest amenities: automatic feeders and waterers, and a top-end heating system. This was neither the first nor the last time Flemming purchased new machinery. New houses and new equipment increased productivity but also dramatically increased the cost of production, a cost that Flemming bore. Flemming was considered a large grower and highly successful. Most farmers were neither as lucky nor as successful. To finance these considerable investments, Flemming worked for wages off the farm—he took on "public work," the derogatory term farmers used to describe wage work. Flemming periodically worked in the Chicopee Textile Mill of Gaines-ville, Georgia, and like thousands of farmers he used his wages to finance poultry industry growth. Farmers moved between farmwork and wage work to finance and sustain broiler raising and to supplement declining farm incomes.[1]

The implementation of the poultry industry's business model—quasi-vertical integration and the feed conversion contract—reshaped the lives and livelihoods of farmers in the Upcountry in dramatic ways. Farmers, as we know, began to invest large sums to purchase broiler housing and machin-ery. They owned the very costly means of production, but lost all control over production and marketing decisions. Increased investment did not—as one might imagine—translate into greater bargaining power. Farmers were investing huge sums of money, mortgaging their homes and land—but this investment did not buy them a seat at the negotiation table. Indeed, this investment translated into greater dependence on integrators. Owning the means of production put farmers in a precarious position. They had single-use buildings, which cost thousands of dollars, and their integrators could drop them at any time for any reason. Integrating firms held monopolies in certain regions, leaving growers with virtually no bargaining power. Farmers could not seek out the best feed conversion rate and contract. They had limited to no options. "The cotton share-cropper," one agricultural economist explained in 1955, "has been replaced by the broiler share-cropper." "Such integration," he maintained, "under the feed dealer's control is essentially monopolistic and therefore increases costs, prevents needed adjustments and results in misallocation of resources and incomes."[2]

Integrator profits increased exponentially, as grower profits declined. To stay in the business, broiler growers began to work off the farm full-time as wageworkers. Between 1950 and 1970, farmers who did not and could not meet integrator demands left the land, and there was a sizeable farm exodus. It is fair to say that the demands made by heads of industry posed a direct threat to the survival of Upcountry Georgia poultry farmers.

By the mid-1950s, farmers with capacities of fifteen hundred to ten thousand birds stood by as contracts went to those who could raise upward of thirteen thousand. In the mid-1950s, only 20 percent of poultry farmers in the Upcountry fell within the large classification. Soon large poultry farmers overran smaller farmers who could not keep pace with integrator demands for increased capacity. As larger poultry farmers produced record amounts of poultry, the rural exodus of marginal farmers grew. Between 1954 and 1969, the number of poultry farms in Hall County fell from 1,138 to 308, a 70 percent drop. Within that same period the rural farm population of Hall County dropped by more than 80 percent, from 11,314 in 1950 to 2,033 in 1970. Other counties watched as comparable numbers of broiler growers fled the region. Leaving the land in the postwar period must have come as a shock to farmers who had weathered the out-migrations that shook the region during World War I, the Depression, and World War II.[3]

A 1956 report showed that, of the 9,200 broiler growers in Georgia, 80 percent were small growers who had to supplement their income either with other farm enterprises or with work off of the farm. "Growers with a capacity of 1,500 to 10,000 birds fear that they cannot compete successfully with single enterprise growers who use mass production methods and have broiler capacities of 13,000 and perhaps eventually 50,000 or more," the report said. "In addition to the fact that even these relatively small growers have considerable investment in broiler production, they are particularly fearful of losing this source of income as they have no alternative enterprise to replace broiler production."[4]

"The big boys has got all the chickens," B. A. Tatum, an Upcountry farmer, told a journalist in 1965. "Just as well not to talk about it," he remarked, adding "they ain't going to turn any of it loose." Tatum had begun raising broilers at war's end and ceased in 1963. His four chicken houses, "collapsing in disuse," sat on what had once been his land. He had lost his land and home. "Poor folks can't do no good now," Tatum concluded. In 1965, broilers accounted for 95 percent of Dawson County's farm income, but this income was by no means evenly distributed. Only farmers raising flocks of ten thousand or more broilers could contract with integrators. Farmers like Tatum fell by the wayside, left with poultry houses and debt. "The steady surrender of littleness to bigness is a central theme in the history of Dawson," the reporter observed, adding, "little poultrymen have lost out to the big feed corporations and big producers." "Bigness" did not equal prosperity for all, in a county where 80 percent of the students qualified for the school lunch program. One of the key "conquerors" of the region, the journalist concluded, was the big corporation.[5]

Part 1. Owning the Means of Production

When integrators began to insist that broiler growers purchase specific houses and machinery, many farmers took pride in the fact that their farms were innovative—new technologies removed some of the drudgery from farm life and increased production and efficiency, and gleaming buildings and machinery seemed like proof that Upcountry farmers were forward-looking, advanced. Poultry farmer Sanford Byers embraced the new equipment, explaining, "I was most modern."[6]

When Upcountry farmers first invested in poultry production during the 1930s and 1940s, the requirements for investment were relatively minimal. Farmers typically built low-cost houses using lumber from their own farms. They used rolled composite material as roofing, and nothing more than dirt as a floor.[7] In 1953, 86 percent of poultry houses still had dirt floors.[8] Farmers made makeshift heaters. According to Welborn, most farmers "had primitive-type heating units." Farmers made heaters out of fifty-five-gallon oil drums, affixed a thermostat to the drum, filled it with coal, and let it burn through the night. Before the use of automatic waterers, women and children filled up jugs in the houses two or three times a day. At that time sweat equity counted more than financial investment. "It was hard work back then," recalled Spurgeon Welborn. "All the catching and loading, and feed was done by hand."[9]

Automatic feeders not only reduced feeding time and diminished much of the tedium of caring for chickens but also prevented the waste of costly animal feed—now that farmers earned their keep through feed conversion contracts, this innovation was essential. Before automation, Flemming lost a portion of the feed as he moved the feed from sacks to bucket to silo. Automation solved the problem of lost feed. Poultry integrators brought farmers feed in bulk in a truck. "You don't touch it," Flemming explained. "[The feed] goes from the truck into the bin, from the bin into the feeder in the chicken house and from that feeder in the chicken house, distributes itself automatically through the chicken house." The automatic feeding devices saved time. Flemming did not have to carry buckets of feed and water to the broilers. With automation and electricity, Flemming checked on his flocks about four times a day, making any necessary adjustments.[10] Integrator Jesse Dixon Jewell confirmed the benefits of new technology. "We have got a house with 10,000 chickens and 1 man could look after them with 1 hour a day." He explained that chicken houses were "equipped with automatic feeders, automatic waterers, and winches for turning the ventilation, and the heat is automatic."[11] We must remember that farmers did not invest in labor-saving devices purely of their own accord. Flemming explained that much of his investment in feeding devices and other improvements came in response to integrators: "Some of it was at my own initiation, but most of it was at their request, see. I bought my first feeders. Didn't nobody ask me to buy them. But I bought them on account of labor saving. Where you'd go down there and in a ten thousand house spend an hour and a half feedin 'em, why, you'd go down there and in fifteen minutes you could put in enough feed to do a half day by emptying it up in your feeders, putting it in your automatic feeder."[12]

Labor-saving machinery allowed farmers to increase production but often meant that farmers worked more, not less, than in years past. Ruby Byers and her husband, who raised eggs, expanded production and invested in new equipment during the 1950s. But the new machinery did not diminish the number of hours the Byers family worked. Ruby Byers referred to the 1950s as "the egg years," a time when she and her family rarely left the farm, even to attend church. The demands of their growing operation, even with new equipment, meant that the Byers family could seldom leave their chickens unattended.[13]

From the 1950s onward, farmers began not only to replace machinery but also to invest in completely new broiler houses. They spent from $6,000 to $12,000 on each broiler house. According to a 1967 government report, growers in Georgia invested $6,400 for buildings, equipment, and five acres of land to establish a ten-thousand-bird broiler enterprise.[14] Many agricultural county agents cheered new construction and celebrated the large cement houses and automated feeding and watering devices that replaced makeshift chicken houses and hand labor. Some county agents welcomed the permanence of new chicken houses as a sign of a prosperous and growing industry. The new buildings, one agricultural agent explained, "are substantial." "Broiler houses," one county agent observed in 1956, "are constructed of concrete blocks and other permanent materials." One farmer built a house measuring forty by four hundred feet that housed twenty-four thousand broilers. Soon houses of this size became the norm, and the only way in which a farmer could continue to stay in the business. Larger houses allowed for increased production, and farmers who in the 1940s raised five thousand broilers raised as many as twenty thousand in 1950. "The trend," one agent explained, "is toward larger flocks and bigger houses with automatic labor-saving equipment."[15]

Oddly, some county agents promised that the "substantial" chicken houses that began to cover the countryside could be used for other purposes in the event of failure in the broiler business. They were mistaken. Broiler houses were highly specialized single-use buildings that could not be adapted to other purposes. Perhaps some county agents were prisoners of or devotees to modernization, and therefore celebrated anything that could be labeled "progress." Or we might also conclude that some county agents wanted desperately for these structures to be multiuse because they were not blind to the fact that the Upcountry was again falling victim to a one-crop economy.

From 1950 onward, integrators demanded both standardized equipment and buildings that housed upward of ten thousand chickens. They instructed growers to abandon newly built houses deemed too small. Growers like Tatum who had assembled makeshift houses, feeding devices, and brooders were forced to abandon homemade devices and invest in costly new machinery. By the mid- to late 1950s, large houses, mechanical means of feeding chickens, and automation were de rigueur.[16] To expedite the cleaning of broiler houses, integrators requested, and eventually mandated, that farmers build houses large enough so that tractors could drive lengthwise through the buildings to clear animal waste and prepare the house for a new flock.[17] They also mandated changes in roofs. Instead of rolled material, the newer houses had aluminum roofs, which helped to keep down the interior temperatures during the summer. Farmers needed to also provide proper ventilation and install "roof ventilators on hip or gable type roofs." Shed roofs were more affordable to build, but they did not allow for proper ventilation.[18]

Farmers began losing autonomy at the very moment when they were increasing investments in their operations. Upcountry farmers struggled to follow integrators' orders, which often seemed unreasonable to those who had spent their lives on farms. Integrator demands struck Sanford Byers as exorbitant. "We have done spent our days in this chicken business," Byers remarked, adding that he knew better than integrators who insisted on costly equipment.[19]

In addition to demanding repeated "upgrades," integrators began to provide "technical supervision." One report explained that a supervisor advised and "sometimes determines the number of broilers a grower may put in a house, the amount of feeding and watering equipment used and many of the disease prevention measures to be taken." As the industry matured, directions became more and more specific, the report explained. "Some dealers prescribe certain standards as to floor space allowed for each bird, amount of feeder and waterer space, and other building features before placing chicks with a grower."[20] Flemming remembered that supervisors came weekly to his farm. He explained that the supervisors "have got full control, and if you don't do a good job you got to hunt you up somebody else," by which he meant that the integrator would drop his contract. Flemming concluded, "you [the broiler grower] are not fully in control all the way through because what that service man asks you to do, why, you do it, see."[21]

Farmers owned the means of production, but integrators barred them from making production decisions. Likewise, under the feed conversion contract, farmers had no say in the marketing and sale of broilers. Spurgeon Welborn lamented the loss of farmer autonomy. He recalled that in the early years, the farmer retained a degree of control, overseeing when and where chickens were sold. As integrators corralled farmers into the contract system, farmers who sought to raise and sell chickens without an integrator found that they could neither purchase chicks and feed nor find markets for broilers. "If you didn't have a contract and [integrators] didn't furnish you feed and chickens," Welborn explained, "you couldn't sell your chickens." Farmers who resisted raising broilers with integrating firms or who resisted the constant demands for new machinery faced bankruptcy.[22]

Integrators locked out farmers who wished to remain independent and had the resources to do so. A farmer might be able to purchase chicks and feed, but finding a market became virtually impossible, and "if you can't sell them and you have to keep them a week or ten days longer than you're supposed to, you could lose all you made," Welborn explained.[23] Flemming echoed Welborn's observations: "If you're not on contract you don't get no chickens any more."[24] During the 1950s, fewer than 2 percent of growers working with feed dealers did business as cash producers.[25] Flemming underscored the impact that contracts had in diminishing farmers' autonomy and forcing them to comply with the integrators' wishes: "There were times when I disagreed with 'em. . . . Now I didn't have no disagreements when I would furnish my own chickens and feed 'em because I was in control of it, see. But on contract basis, they ask you to do these things and you almost got to do 'em to stay in business with 'em."[26]

Many farmers and observers of the industry feared that poultry farmers were becoming hired hands. Even David W. Brooks, the head of Gold Kist, a cooperatively owned integrating firm, echoed the complaints of farmers and feared that poultry farmers would be reduced to wage laborers. In 1957, Brooks concluded that "the grower was becoming more or less a hired man for the feed dealers and the feed manufacturers." Brooks warned that soon landowning farmers would "have no more control over [their] future than any laborer in an industrial plant." Confirming Brooks's fears, government studies of the contract system found that most farmers had "little voice" in production and marketing. In part, Brooks and others were correct in likening the position of farmers to wage laborers, but unlike wage laborers, poultry farmers were independent contractors. As such, they were not entitled to any of the protections and rights accorded wage laborers: a minimum wage, workers' compensation, and social security. Poultry farmers invested in poultry houses but were never guaranteed a steady flow of flocks. The integrator could simply cut the number of flocks a farmer received, leaving the farmer with no source of income and a sizeable debt.[27]

Farmers had become hired hands in the very factories that they owned. In the mid-1960s, another set of researchers found that growers in Georgia earned returns of fifty-three cents an hour. Twenty-five years later, researchers pointedly identified growers as wage laborers. "These farmers have in essence, become hired help for corporations," they wrote, "while the contractors benefit from not having to pay minimum wage to the growers."[28] Interviewed in 1987, Spurgeon Welborn described the farmers' predicament in even starker terms: "The poultry producers now are tenants," he said. "They are paid so much a dozen for producing eggs, and they're paid so much a pound for producing broilers. They furnish the labor and the house. The big companies furnish the chickens and the feed, and they're really just tenant people that are hired."[29]

Celebratory accounts of the broiler industry touted the 1950s as the good years, a period of growth and expansion. Indeed, the new buildings and state-of-the-art machinery seemingly confirmed that the industry was profitable for everyone. The new buildings and machinery were viewed as markers of innovation and progress. In truth, the buildings were monuments to or markers of indebtedness, inequality, and lost autonomy, not evidence of wealth and progress.

Part 2. "They Can Ruin You"

Well before quasi-vertical integration and the feed conversion contract were fully in place, farmers had begun to lose ground. When they tallied their incomes in the years immediately following the end of the war, they began to realize that growing gross income did not translate into growing net income. The rising cost of production impinged on profit.[30] Many poultry farmers were just breaking even, a county agent reported, adding that most farmers planned to stay in the business hoping that net profits would rise.[31] As early as 1947 there were stark indicators that farmers were losing ground. In a 1947 report, one county agent recorded that farmers were losing four cents a pound on broilers. It cost farmers twenty-eight cents to produce a pound of broiler meat, but the market price was twenty-four cents a pound.[32] Throughout the Upcountry, farmers' net incomes declined.[33] "The economic position of Georgia farm families," one farm agent reported in 1949, "is not as favorable as it was a few years ago."[34] Immediately after the war, H. A. Maxey, Cherokee County's extension agent, wrote, "One of our greatest problems is to keep people farming and to keep them from going too deep into the poultry business."[35]

From 1950 onward, gross profits and production increased, but net profits to poultry farmers shrank. Investment and production costs cut deeply into farmers' proceeds. In 1950 Georgia poultry farmers earned on average a net farm income of $978; ten years later they earned $777. Over this same ten-year period operating expenses more than doubled. Conditions among farmers and consolidation among integrators prompted a congressional investigation that confirmed that farmers had increased production but earned less and less money each year.[36]

Farmers were working harder, investing more money, and producing more chickens, but their profits continued to decline. In 1957, during congressional hearings investigating problems within the poultry industry, Jesse D. Jewell testified that, prior to 1953, poultry farmers earned between $150 and $200 for every one thousand chickens raised. In 1957, they earned $10 for every one thousand chickens raised. Prior to 1953 the average farmer raised a flock of about two thousand chickens, in 1957 four thousand chickens. So in 1957 farmers who raised twice as many chickens as they had four years earlier earned less than half of what they had earned in 1953.[37]

Integrators, not farmers, reaped the rewards of increased efficiency. One government study concluded that technological advances increased the initial cost to farmers who established chicken farms. Farmers were paying more to enter the poultry business. As integrators like Jewell demanded newer equipment, farmers, according to a government report, sought out credit in "amounts that had previously been neither anticipated nor needed." Farmers took on debt that researchers deemed to be financially risky.[38]

Poultry farmers and integrators had reached an unequal bargain. A 1976 government report explained the farmers' predicament. Integrators had "shift[ed] to the grower many of the costs and risks of the business. The grower is often saddled with heavy debts in the form of mortgages on his land and buildings and is faced with the choice of accepting the low returns offered by the contractors or losing his farm."[39] Rising gross profits, the proliferation of wage work in poultry processing plants and related industries, and the continuous new construction of broiler houses concealed the fact that the heads of the industry received a disproportionate share of the proceeds, as they shifted financial burdens and risk to farmers.

Integrators named the price to be paid for growing broilers; they did not negotiate with growers over the feed conversion ratio. If farmers disagreed or objected, integrators simply refused to contract with them; farmers were then left with large debts for buildings and equipment and empty broiler houses. Not surprisingly, Jewell, like most integrators, refused to take any responsibility for the low profits farmers earned. His business practices, he ardently maintained, did not contribute to growing poverty in Upcountry Georgia. "The market, supply and demand," Jewell maintained, "governs what [farmers] make."[40] It was not the free market but integrators who determined farmers' profits. Farmers' earnings bore little to no relationship to the market price of poultry. Integrators held monopolies in specific areas. Therefore, an integrator had tremendous power, because farmers could not negotiate with multiple firms to get a better feed conversion rate.[41] Farmers, in most cases, could contract with only one integrator and thus had no ability to negotiate rates.

Over the course of the 1950s, there was a growing discrepancy between farmers' net and gross profits and an even greater discrepancy between the profits of integrators and those of farmers. Income tax, social security, and production costs cut deeply into farmers' profits. County agents charged with aiding farmers worried about conditions among poultry farmers and in the poultry industry more generally. In 1954, Cherokee County produced roughly twenty million chickens on twelve hundred farms and was the largest poultry producing county in the state. Even so, low returns to farmers persisted. Farmers had few if any alternatives to poultry farming, and the county agent maintained that chicken was "our only hope." "The broiler industry has grown so big it is not very profitable for anyone," a county agent explained, adding, "it is still the best thing the farm people in this county can do."[42]

As a remedy for low returns, county agents began to counsel greater efficiency and recommended that farmers buy the newest equipment, echoing the demands integrating firms were already mandating. "An increase in efficiency in farming," explained one agent, "is a must if farmers are to continue to receive a desired standard of living from farming."[43] Another agent explained that the slogan for 1954 was "Increase Farm Profits with Quality and Efficiency."[44] As county agents counseled farmers to increase efficiency and purchase new equipment, they effectively encouraged farmers to take on more debt.

Agricultural agents followed a course not unlike that of their predecessors, who in the first decades of the twentieth century had counseled cotton farmers to increase efficiency. Fertilizers, engineered seeds, pesticides, and machinery increased cotton yields, but they cost a great deal of money and only compounded the glutted cotton market that undermined farmers' returns. Likewise, purchasing new poultry houses and equipment in the name of efficiency and increased production did not translate into higher profits for broiler growers. Soon, many farmers began footing the bill with money earned from wage work off the farm. The cost of growing broilers increased, as grower autonomy and profit declined.[45] As early as 1951, a study that analyzed financing arrangements between farmers and integrators concluded that farmers were sacrificing their financial independence. According to the study, "a review of the various financing plans shows that under some plans the grower merely becomes a laborer, with very little voice in production and marketing."[46] Broiler grower Welborn explained the precarious position contract farmers were in by the late 1980s. A farmer, he explained, "spend[s] a hundred thousand dollars on houses pretty quick, and if [integrators] decide they don't want no more to do with you, you've had it. They can ruin you."[47]

Part 3. Public Work

The broiler growers who remained in the Upcountry took on second jobs to supplement their farm income and pay for costly new poultry raising equipment. In the early and mid-1950s, county agents and agricultural researchers began to document a growing trend. Meager profits and rising production costs pushed poultry farmers to seek supplemental sources of income. "The farm population is caught in a squeeze of rising costs and lower prices for goods sold," a county agent observed in 1954, adding, "many farmers are already working part or full time off the farm."[48] The agent feared that conditions for farmers would only worsen. Two years later, his fears were confirmed, as rising costs, coupled with lower prices, continued to undermine the profits of farmers. The agent explained, "Farmers are being forced in increasing numbers to seek supplemental income from sources outside the farm."[49] In 1956, only 14.5 percent of poultry farmers in Upcountry Georgia worked off-farm for wages. By 1962, over 50 percent of Upcountry Georgia poultry farmers or their family members took jobs off the farm to supplement meager farm incomes and to meet the demands of poultry industry leaders. Georgia poultry farmers and their families, earning average net farm incomes of $777, made on average $1,459 from wage work off the farm. By 1970, over 70 percent of Upcountry chicken farmers supplemented farm incomes with wage work.[50]

Poultry farmers found wage work in a range of industries related to poultry production. In Hall County alone in 1955, there were eight poultry processing plants, six poultry storage facilities, twenty-nine hatcheries, one fertilizer plant, one crate factory, and one pet food plant. Hall County also was home to one air conditioner and compressor plant, a marker of the rise of the Sunbelt. The textile industry continued to employ farmers in Hall County, and in 1955 two cotton mills and one hosiery plant hired farmers. Beyond Hall County, farmers found work at General Motors and in the growing military-industrial complex. A Lockheed plant was roughly sixty miles west of Hall County. Nearby Cherokee County was home to the Purina Research Farm, the largest commercial broiler research farm in the United States. In Canton, Cherokee County's seat, a cotton mill employed fourteen hundred workers, many of whom continued to operate their farms.[51]

Jewell and other industry leaders cast the growing practice of combining wage work and poultry production as a choice, a means of acquiring consumer goods, and a boon. This was an absurd claim. "The womenfolks live in town and work at the different factories or dressing plants," he explained, adding, "they seem to get along better now than they ever have before." Not only did farm women live in town, but their husbands were able to acquire automobiles. Farmers "are able now to get jobs in industry, they have got better roads, and they all have got cars," Jewell cheerfully explained. Jewell went so far as to claim that farmers who worked off the farm supplemented wage work with their income from the chicken business, rather than the converse.[52]

Here was a businessman—Jewell—who both touted his business as the region's savoir and then shifted his narrative. Farmers, according to Jewell, loved wage work off the farm and supplemented that wage work with broilers. According to Jewell, broilers had become just a sideline business, a minor source of income. Jewell's shifting narrative is disturbing at best—a means of abdicating responsibility. In that same 1952 interview, Jewell stated that he provided "an important, profitable, sideline for 600 farmers in the area who fatten[ed] the chickens for him."[53] It might be fair to say that Jewell was performing a sleight of hand. He offered folks a sideline business; he was not in fact saving farms, farmers, and rural life. Growers, he stated, raised chicken as a secondary source of income—this, of course, was a bald-faced lie.[54]

In addition to the record numbers of cars lining Gainesville's streets, Jewell marveled that wage work and poultry production had added consumer goods to Upcountry Georgia's landscape in other ways. "There are more people with more money," Jewell asserted, adding that, for the first time, Upcountry Georgians were able to purchase good clothes. Other observers similarly credited the poultry industry with providing the "adjuncts of a good living." One county agent tallied the range of appliances farmers purchased: stoves, refrigerators, washing machines, irons, radios, and televisions. The flood of consumer goods that Upcountry Georgians purchased helped to reify the notion that the poultry industry was the region's savior. It put money in the pockets of poor farmers and brought all the accoutrements of modern middle-class life.[55]

Farmers who juggled wage work and farmwork took a more sober view. Jewell and other integrators cast wage work as something farm families joyfully embraced. For farmers "public work" was simply a necessity. To supplement his farm income, Flemming worked at the hosiery plant in Gainesville, Georgia, at night and tended his chickens during the day. "I never had been fastened up in a building before," Flemming remarked. "I had a space about four foot wide and about twenty-one feet that I walked eight and ten hours a day." The experience left a lasting impression on Flemming, who vividly remembered the dimensions of his workspace and the tedium of his work.[56]

As farmers bemoaned factory conditions, county and home demonstration agents worried about the effect of wage work on farm and home life. County agent Maxey worried about the trend of farmers seeking wage work. As growing numbers of farmers went to work in factories, one home demonstration agent feared that the farm family would deteriorate. The advent of wage work presented a problem, the agent noted, "especially when a father may work on one shift and the mother on another." Farmers and county agents worried over the hours farm families worked both on and off the farm. But the number of farmers seeking public work continued to grow.[57]

Wage work supported the growth of the poultry industry to the overwhelming disadvantage of the farmers themselves. In contrast to their mothers and fathers, who had worked for stints in the nearby Chicopee textile mills to supplement their cotton farming incomes, more than 50 percent of poultry farmers worked full-time, supplementing their farm incomes with "public work" during the 1960s. Poultry farmers were different from their predecessors in another crucial respect. Unlike the cotton farmers before them, Georgia poultry farmers had mortgaged their homes and had incurred an average of $10,000 of debt to build and maintain poultry houses. Meeting the demands of poultry firms, farmers subsidized the growth of the poultry industry by investing in the newest technologies that facilitated new economies of scale. Farmers' wage work to cover these investments made a mockery of the claim that poultry production single-handedly preserved Upcountry Georgia farms.[58]

෴

Farmers in Upcountry Georgia found that they had replaced one form of one-crop agriculture with another, helping to create a system that reinforced the widespread poverty and an intractable class system long associated with cotton planting. Flemming reflected on the conflicting ways that farmers made sense of their predicament. He found wage work constricting and poultry production increasingly onerous financially. Even so, Flemming insisted, "the chicken business was a blessing to north Georgia." But it is important to note that Flemming was part of the minority of growers who actually remained solvent and succeeded. He began raising chickens in 1939, worked for wages in a textile mill to support his farm, and purchased new poultry farming machinery six times over roughly forty years. He was considered successful, even an exemplary grower. In 1951, *Broiler Growing* featured Flemming on its front page, showcasing his housing and machinery. The article did not explain that he worked for wages off the farm to pay for all this machinery he was forced to buy. Not until 1981 did Flemming leave poultry farming, after refusing to commit the $24,000 to $30,000 his integrator had demanded that he invest in new machinery.[59] The majority of Upcountry farmers had a shorter run in poultry farming, as they failed and left the land. It is true that there also were factors pulling them off the land. More industrial jobs were available, and education through the G.I. Bill had opened new opportunities to veterans from Upcountry Georgia. Yet poultry farming played a significant role in pushing farmers off of the land.

CHAPTER 5

From Public Nuisance to Toxic Waste, 1940–1990

IN 1950, a journalist for *Colliers* magazine wrote a laudatory article about Upcountry Georgia's broiler industry, describing with great élan the smell that pervaded the city. "When the wind is right," the journalist wrote, the "citizens of Gainesville, Georgia, including occupants of the expensive new homes . . . get a whiff of it." The smell was not unpleasant, "a mere soupçon of scent . . . unmistakable to anyone who in youth has ever done chores around a henhouse." In the article, aptly titled "Chickens in the Wind," the journalist claimed that the residents of Gainesville loved the smell. "They throw back their heads and breathe deeply—the richer they are, the more rapturously they inhale."[1] The journalist accurately reported a pervasive odor. It was untrue, however, that all Gainesville residents loved that smell. The "mere soupçon of scent" had become all-pervasive and foul, if not revolting. As the industry grew, piles of dead chickens, manure, and offal could be found on farms, in fields, beside processing plants, and in municipal sewage systems. In addition to poultry manure, the poultry industry unleashed pesticides, hormones, and antibiotics into the air and water. In the space of less than twenty years, broiler production had increased by roughly 200 percent.[2] Georgia farmers produced half a million chickens in 1935, 88.6 million in 1951.[3] Over time, the growing industry produced incalculable amounts of waste. From the end of World War II onward, poultry production polluted the soil, air, and water. Rapid growth and lack of regulation meant that waste—in a range of forms—spread unchecked. In short, poultry integrating firms treated the land and its inhabitants with reckless disregard.

Part 1. The Problem of Growth

From World War II onward, the scale of chicken farming and the waste produced by chickens, rendering and processing plants, and feed mills ballooned. Manure produced by a flock of fifty thousand chickens could not easily be used to fertilize cotton crops. Hundreds of dead chickens could not be buried or burned without complaints from neighbors. Plants processing ten thousand or more birds per day could not hide the fact that they were dumping excessive amounts of runoff, waste water, and offal into municipal sewage systems. The broiler industry had grown very large very quickly, and Upcountry residents were now living with the consequences.

As farmers expanded the size of their flocks, the disposal of the chicken manure produced by flocks of fifty thousand or more birds became a pressing problem. Initially, farmers had used the manure to fertilize their own cotton fields and save on commercial fertilizers.[4] Velva Blackstock, an Upcountry chicken farmer, remarked that chicken waste was a good fertilizer, but too much manure could harm a field crop.[5] Blackstock remembered, "My daddy used to say you'd better not put it on there in the year that you want it to produce, because unless you get a certain amount of rainfall, it can cause your crop to burn."[6] During the mid-1950s, farmers still spread the manure on surrounding fields. In 1956, one government source estimated that almost 90 percent of growers used chicken manure on their farms. According to the report, farmers scattered the waste on their fields.[7] But as farmers ceased growing cotton altogether, the abundance of chicken manure became a burden, not a remedy for soil infertility.

Chicken waste soon became a problem not only in the Upcountry but also in new broiler-producing areas throughout the South. In one southern town, a chicken farmer, according to nearby residents, "allowed the manure droppings from the thousands of chickens to accumulate and stand for days and weeks at a time." Residents observed that the farmer "negligently allow[ed] water to mix with . . . chickens' droppings and manure," creating a breeding ground for flies. The farmer allowed piles of manure to grow several feet high, left dead birds lying around his farm, and periodically burned dead chickens. The "odor of the manure and droppings . . . mingled with the odor of dead chickens and burnt feathers" so nauseated neighbors that they ceased going outdoors.[8]

As integrators pushed farmers to expand capacity, they not only intensified the problem of waste but also created conditions that resulted in higher rates of chicken mortality. In 1952, a poultry respiratory illness killed thousands of birds, in some cases entire flocks.[9] New poultry diseases were an outgrowth of large-scale production and crowding in poultry houses. Integrators, though they owned the birds, charged farmers with the task and cost of disposing of chicken carcasses. Lacking a proper system to dispose of birds and any knowledge of how to manage an unprecedented amount of waste, most farmers tossed dead chickens in heaps behind chicken houses. County agents counseled the "prompt disposal of dead birds in a disposal pit" in order to control disease, although it is unclear how agents determined that this was the best way of handling the problem. After all, these disposal pits directly contaminated watersheds, streams, and rivers. To be sure, farmers, county agents, and integrators were not equipped to cope with an environmental problem of this scale. In 1952 only one poultry farmer in all of Cherokee County had a disposal pit for dead chickens.[10]

While the chicken houses grew in size and number, so too did the processing and rendering plants. As was the case with chicken manure, integrators had not made provisions for the new form of waste—the tons of offal that remained behind as broilers were shipped to market. Government regulations during World War II revolutionized the processing and distribution of broilers. As a result, firms such as Jewell's Poultryland Inc. and others no longer sent live birds or even New York–dressed birds to markets in the North and South. Consumers now expected and purchased dead and cleaned chickens that bore little resemblance to the spring chicken. Chickens now left the Upcountry in very much the same way that hogs and beef left packing housing in Chicago. Workers in processing plants created "a compact product," ready for sale in America's supermarkets.[11] Producing this "compact product" meant that tons of waste was left behind in the Upcountry.

Industry leaders had not developed effective mechanisms for managing the growing waste, and the problem only amplified. In 1943, L. C. Rew arrived in Gainesville to assume his position as the county agent, and he found workers from processing plants carrying chicken heads, feet, and intestines and scattering them on fields. Feathers, blood, and inedible offal were washed into the sewage system, and piles of dead animals generated odors and swarms of flies.[12]

Industry leaders used waste as chicken feed. In 1944, Jesse Dixon Jewell opened a rendering plant within the Gainesville's city limits to manufacture feed and fertilizer from chicken and other animal parts. At a congressional hearing, he explained how his plant turned waste into animal feed, revealing his indifference to the problems that might arise from feeding chickens the rendering plants' detritus: "We take the offal, the heads, the feet, the viscera, and make a meat scrap out of them which is a very desirable meat scrap. We take the feathers and dry the feathers and that is a very good source of protein, and that all goes back into the chicken feed."[13]

Part 2. Nuisance Suits

Even though their understanding of the full implications of the environmental impact of the poultry industry remained limited, residents and farmers registered their concerns. As farmers increased the size of chicken houses and expanded their flocks, residents in rural areas and small towns began filing complaints, specifically public nuisance suits, and petitioned courts to force farmers to cease operations. Neither modern environmental regulation nor environmental law existed yet, so residents filed public nuisance suits. Nuisance is one of the oldest common-law offenses. A nuisance is something—noise, smells, waste—that disturbs a property owner's right of "quiet enjoyment" of his or her property. Accordingly, plaintiffs responding to the excesses of the poultry industry charged that a chicken house or rendering plant was a public nuisance that prevented residents from enjoying their property and potentially depressed property values. As one legal historian explains, "The statutory form of modern environmental law is built on the backbone of the common law of nuisance."[14]

In 1944, residents who lived near one of Jewell's rendering plants brought suit against Poultryland Inc., charging that it was a public nuisance and petitioned for its closure. The list of offenses was long. According to court documents the plant emitted "vile, offensive, and obnoxious odors, gas, and vapors," which one could smell for a mile and half away. The "obnoxious odors" made houses and businesses uninhabitable. Residents charged that the smell was so overpowering that people became nauseated and were "unable to eat their food with any relish." In the proceedings, the petitioners claimed that "the pillows, mattresses, and bed clothing in their homes . . . [were] so contaminated by the said odors" that residents could not sleep. The plant was not merely a source of bad odors. Dead horses and mules were left outside the plant to rot. When the plant began operations, Jewell disposed of waste in the sewer system that served the surrounding residential community. The sewers were overtaxed, and the waste soon bubbled up into the streets of Gainesville. Court documents reported that waste from the plant "poured out through the manholes." When the city replaced perforated manhole covers with solid ones, the "noxious vapors and gases" backed up into the sewer system and into residents' kitchens and bathrooms. To address the problem, Jewell ceased depositing waste into the sewer system, but the plant's smoke stack continued to "belch and pour out" polluted air.[15]

In 1959, when a farmer expanded his flock to forty thousand broilers, residents of one rural town in Arkansas filed a suit charging that the growing operation caused their property values to depreciate. Neighbors complained that the farmer's chicken operation smelled bad and attracted flies and that his chickens "made loud and distracting noises at all hours of the day and night." Neighbors charged that dust and litter from the chicken house saturated the air and reduced the market value of their property.[16] Most petitions and complaints pointed to the same set of problems: obnoxious odors, loud noises or a "constant cackle," piles of manure, piles of dead birds, burning of birds, and swarms of flies.[17]

As the poultry industry grew in the 1950s and 1960s, conditions like those found in Jewell's plant persisted and worsened throughout the South, and residents and neighbors continuously filed nuisance suits in an effort to force plants to close. In Arkansas, residents complained that a rendering plant "emitted noxious odors and attracted flies." They charged that the plant left offal in uncovered steel barrels where it became infested with maggots.[18] Residents in rural Maryland charged that a rendering plant produced a "shocking and nauseating stench and odor which permeate[d] the surrounding atmosphere for more than a mile." Those who lived in the vicinity of the plant suffered "throat irritations, severe headaches, loss of appetite, nausea," and vomiting.[19] Even so, rendering and processing plants multiplied and grew. As proof of industry growth and the problem of waste, one Mississippi rendering plant processed 150,000 pounds of offal in 1954, 75 million pounds in 1962.[20]

Plaintiffs filed nuisance suits because they had no other recourse, but their arguments tended to fail. The power of the plaintiffs' arguments was limited to an emphasis on property rights because there was not yet evidence that the waste was toxic and a threat to human health. In the 1950s and 1960s, Americans were only beginning to comprehend industrial pollution, and it is likely that most people did not imagine that farms could pollute in the same way that a factory could. The American farm, no matter how it operated, continued to conjure pastoral images as the bucolic foil to industrialization. These persistent myths impeded Americans' ability to fully appreciate the damage that farming could cause to the land and to humans. It is understandable, then, that in the nuisance suits filed in the 1950s and 1960s there is a conspicuous absence of claims that the poultry industry caused health hazards or illness. When plaintiffs did charge that a broiler house, processing plant, or feed mill was a health threat or source of water contamination, suits were often if not always thrown out for lack of evidence.[21]

Judges found themselves in a precarious position. They could shut down farms—and inhibit economic growth in regions badly in need of jobs—or they could allow the broiler industry to carry on. In an Arkansas case, the judge ruled that neighbors could not force the farmer to close his operation. Instead, the judge ordered that the farmer not leave manure on his premises for more than twenty-four hours and required that the farmer bury, not burn, dead chickens. In his opinion, the judge emphasized that poultry raising was lawful and was becoming central to the economy.[22] To be sure, it is quite odd that a judge—not a poultry or environmental scientist—determined how best to manage what we now know to be toxic waste.

It would take decades for scientists to determine that indeed this waste threatened human health. Even without this scientific knowledge, those who lived adjacent to broiler houses and processing plants intuitively knew that something was awry. Residents filed suits in which they charged that the manure that made its way into streams and rivers was "unhealthful." Likewise, they maintained that the stench from piles of dead birds and the burning of dead birds sickened their children. But they lacked scientific evidence to prove the correlation between poultry waste and health problems. It was not until quite late in the twentieth century that scientific studies would prove that indeed poultry waste is toxic. Furthermore, it was not until the 1990s that the federal government would begin to regulate agricultural pollution. Thus, it is not remarkable that this pollution did not provoke concern on the part of the local and state officials. After all, the broiler industry was the region's savior, and Gainesville was the self-proclaimed poultry capital of the world. Officials had no desire to place restrictions on industry growth.

Part 3. What We Now Know

The many phases and components of poultry production are harmful to human health and the environment. Poultry waste contaminates watersheds.[23] Chicken feed has been linked to respiratory illnesses among farmworkers. Subtherapeutic antibiotics—used to fatten chickens at a record pace—find their way into the American diet and are now linked to the epidemic of obesity.[24]

Regulations established in response to new scientific knowledge have been toothless in relationship to agribusiness. The Clean Water Act, from which, revealingly, the farm lobby pushed to be exempted, was limited in its power to address the pollution unleashed by industrial agriculture. The Clean Water Act of 1972 largely primarily regulates "chemicals or contaminants that move through pipes or ditches." It does not "apply to waste that is sprayed on a field and seeps into groundwater."[25] In the 1970s, the Clean Water Act began to regulate "point source pollution": pollution associated with factories generally coming from a discrete source such as a pipe or ditch.[26] The poultry industry, like most forms of agribusiness, escaped regulation, and the legislation did little to stem growing waste lagoons and the poultry industry's risky environmental practices. To this day "runoff from all but the largest farms is essentially unregulated by many of the federal laws intended to prevent pollution and protect drinking water sources." The Environmental Protection Agency (EPA) did not begin to regulate non–point source pollution—the runoff of animal waste, pesticides, hormones, and antibiotics—until the mid-1990s.[27]

As environmental law has become more stringent and as the poultry industry has continued to expand, poultry integrators have benefited from the status of poultry growers, who, under the conditions of their agreements, are independent contractors. Therefore, growers, who are often small farmers with enormous debt, bear sole responsibility for environmental violations while their integrators—companies like Tyson, Smithfield, and Perdue—evade liability. Once again, the success of integrators rests on the backs of growers whose status as "independent" is a fallacy.[28] The cost of disposing of used litter became onerous and also contributed to the growing environmental problems associated with the industry.[29] And currently farmers and integrators are battling over who is responsible for the high cost of disposing of chicken manure.[30]

Large agricultural firms continue to pollute on an ever larger scale with near impunity. The EPA has reported that today agricultural runoff is the largest source of water pollution in the nation's rivers and streams.[31] "According to a 2009 report by a state-federal task force, cattle, poultry and pigs generate 5½ times more excrement annually than the entire human population of America. The task force found pollution caused by manure and fertilizer has grown dramatically over the past fifty years and efforts to control the damage by state and federal regulators have been 'collectively inadequate.'"[32] In 2011, an EPA inspector found that nearly three-fourths of state inspections in Georgia of forty-eight large farms known as "concentrated animal feeding operations" were faulty or incomplete.[33] The list of violations of environmental regulations, environmental disasters, health scares, and deaths wrought by American agribusiness continues to grow. The problem of pollution is severe. Regulations are weak and go unenforced.

ᔕ

By the 1960s, it was clear that the massive chicken farms and plants had taken a significant toll on the local environment. But the industry's comfortable relationship with the USDA and other federal agencies allowed it to sidestep responsibility for most of the damage it wrought. Even after the establishment of the Environmental Protection Agency, the poultry industry, casting itself as the champion of the small farmer, escaped regulation and remains today the least regulated major industry in the nation.

EPILOGUE

As Americans face the consequences of industrial chicken production, U.S. corporations are already well on their to way to exporting not simply American-grown and processed chicken but also the business model and production practices that have indelibly harmed American farmers, workers, consumers, and the environment.

The varied production chains and the reach of global corporations can stretch the imagination. If you find yourself in Tokyo eating yakitori, skewered grilled chicken meat (dark meat, not white meat), it is quite likely that your chicken was raised in the American South. Its breast meat was removed and processed in the American South, and its dark meat was then sent to Mexico for the labor-intensive process of deboning, then it made its way to Japan.[1] And if you find yourself in Liberia, Equatorial Guinea, or the Ivory Coast and are eating pork, your pork was likely produced in Eastern Europe, possibly in Romania by contract farmers working for Smithfield Farms, a Virginia-based company and one of the largest U.S. pork producers.[2]

In 2007, Tyson Foods said it expected revenue in its international business to rise to $5 billion in 2010, from $3 billion in 2007. Tyson explained that it planned to expand trade with China and production in Brazil and Mexico.[3] In 1988, Tyson created Corporacion Citra, a $42 million joint venture with Trasgo Group (Mexico), C. Itoh & Co. (Japanese trading company, Tokyo), and Banco Nacional de Mexico. Citra consisted of new broiler breeder farms, hatcheries, and deboning plants in Mexico and was established to process and market broilers produced in Mexico and the United States.[4] Citra "adds value" to Tyson-produced poultry, meaning that unprocessed dark meat is shipped from Tyson plants in the United States and processed by Mexican workers in Citra's processing plant, which employs fifty-five hundred workers. As of 1988, the Japanese people consumed 375 million pounds of yakitori per year. With this joint venture, Tyson was attempting to serve the Japanese market.[5]

Southern agribusiness began to export poultry production technology in the 1960s to Western Europe and Japan. Thailand modernized its industry in the 1970s, and in the 1990s Cargill created Sun Valley Thailand, a joint venture with Nippon Meat Packers of Japan to supply Japan with chicken.[6]

Brazil modernized its industry in 1970s as well. And Cargill now produces and processes chicken in Brazil for sale in the Middle East.[7]

Over the course of the past ten years, Smithfield Foods has made significant inroads to pork production in Eastern Europe. The company has essentially transplanted the poultry industry's quasi-vertical integration model in Eastern Europe; it has built and acquired meat processing plants, feed mills, and cold storage facilities. And according to a company spokesman, the company supports "networks of independent farmers that are contracted to shelter and feed pigs to market weights." In North Carolina, Smithfield followed the poultry business model. As the company grew, the number of hog farmers declined by 90 percent: 667,000 hog farms in 1980, 67,000 in 2005. North Carolina placed restrictions on the building on hog farms (especially after the overflow of hog farm lagoons during Hurricane Floyd in 1999). Facing restricted growth in the South, Smithfield headed to Eastern Europe and began to benefit from EU farm subsidies. According to the *New York Times*, "In Romania, the number of hog farmers has declined 90 percent—to 52,100 in 2007 from 477,030 in 2003—according to European Union statistics, with ex-farmers, overwhelmed by Smithfield's lower prices, often emigrating or shifting to construction. In their place, the company employs or contracts with about 900 people and buys grain from about 100 farmers." The numbers are similar in Poland, where the number of hog farmers declined by 56 percent between 1996 and 2008.[8]

The triumph of agribusiness has come with enormous human and environmental costs. Food reaches into the lives of all Americans, and the country faces tainted food supplies, ever mounting and illogical farm subsidies, an epidemic of obesity, contaminated water supplies, and persistent labor violations and workplace injuries.

At the close of the twentieth century, agribusiness leaders looked abroad in search of cheap labor, relaxed if not nonexistent environmental regulations, and regions where quasi-vertical integration could be easily applied. When Smithfield Foods headed to Eastern Europe in the early 2000s, the *New York Times* reported that "the upheaval ranks among the Continent's biggest agricultural transformations." An agricultural transformation that took decades in the South was now compressed into four years. We can only begin to imagine the economic, social, and environmental impacts of the industry.

NOTES

Introduction

1. "Too Few Farmers Left to Count, Agency Says," *New York Times*, October 10, 1993, accessed December 20, 2012, http://www.nytimes.com/1993/10/10/us/too-few-farmers-left -to-count-agency-says.html.

2. Charles Duhigg, "Health Ills Abound as Farm Runoff Fouls Wells," *New York Times*, September 17, 2009, accessed October 15, 2012, http://www.nytimes.com/2009/09/18/us /18dairy.html?pagewanted=all.

3. Mark Bittman, "Pesticides: Now More Than Ever," *New York Times*, December 11, 2012, accessed December 12, 2012, http://opinionator.blogs.nytimes.com/2012/12/11/pesti cides-now-more-than-ever/?src=me&ref=general.

4. Regional histories on the industrialization of southern agriculture include Pete Daniel, *Breaking the Land: The Transformation of Cotton, Tobacco, and Rice Cultures since 1880* (Urbana: University of Illinois Press, 1985); Gilbert C. Fite, *Cotton Fields No More: Southern Agriculture, 1865–1980* (Lexington: University Press of Kentucky, 1984); and Jack Temple Kirby, *Rural Worlds Lost: The American South, 1920–1960* (Baton Rouge: Louisiana State University Press, 1987). Important industry histories include Steve Striffler, *Chicken: The Dangerous Transformation of America's Favorite Food* (New Haven, Conn.: Yale University Press, 2005) and Roger Horowitz, *Putting Meat on the American Table: Taste, Technology, Transformation* (Baltimore, Md.: Johns Hopkins University Press, 2005). The work of journalists is indispensable to the study of agribusiness. See Michael Pollan, *The Omnivore's Dilemma: A Natural History of Four Meals* (New York: Penguin, 2007) and Eric Schlosser, *Fast Food Nation: The Dark Side of the All-American Meal* (Boston: Houghton Mifflin, 2001). On the history and politics of the term "agribusiness," see Shane Hamilton, "Agribusiness, the Family Farm, and the Politics of Technological Determinism in the Post–World War II United States," *Technology and Culture* 55, no. 3 (July 2014): 560–90.

5. On the marginalization of white farm women in the South, see Lu Ann Jones, *Mamma Learned Us to Work: Farm Women in the New South* (Chapel Hill: University of North Carolina Press, 2002). On the plight of African American agricultural workers, see Pete Daniel, *Dispossession: Discrimination against African American Farmers in the Age of Civil Rights* (Chapel Hill: University of North Carolina Press, 2013); Adrienne Petty, *Standing Their Ground: Small Farmers in North Carolina since the Civil War* (New York: Oxford University Press, 2013); and LaGuana Gray, *We Just Keep Running the Line: Black Southern Women and the Poultry Processing Industry* (Baton Rouge: Louisiana State University Press, 2014).

Chapter 1. From Cotton to Chicken, 1914–1939

1. J. T. Holleman, "Is the South in the Grip of a Cotton Oligarchy?," pamphlet, 9– 10, quoted in Rupert Bayless Vance, *Human Factors in Cotton Culture: A Study in the*

Social Geography of the American South (Chapel Hill: University of North Carolina Press, 1929), 188.

2. Willard Range, *A Century of Georgia Agriculture, 1850–1950* (Athens: University of Georgia Press, 1954), 90–166; Wallace Hugh Warren, "Progress and Its Discontents: The Transformation of the Georgia Foothills, 1920–1970" (master's thesis, University of Georgia, 1997). For growth in cotton production after the Civil War throughout the South, see Harold D. Woodman, *King Cotton and His Retainers: Financing and Marketing the Cotton Crop of the South, 1800–1925* (Washington, D.C.: Beard Books, 2000); Gavin Wright, *Old South, New South: Revolutions in the Southern Economy since the Civil War* (Baton Rouge: Louisiana State University Press, 1996).

3. Holleman, "Is the South in the Grip?"

4. U.S. Department of Agriculture, *Economic Needs of Farm Women* (Washington, D.C.: Government Printing Office, 1915), 50.

5. Andrew M. Soule, "The Need of Adequate Rural Leadership in Georgia" (speech, ca. 1923), Andrew M. Soule Papers, Hargrett Rare Book and Manuscript Library, University of Georgia, Athens, box 21.

6. Steven Hahn, *The Roots of Southern Populism: Yeoman Farmers and the Transformation of the Georgia Upcountry, 1850–1890* (New York: Oxford University Press, 1983), 9.

7. Ibid., 4.

8. Ibid., 152.

9. Ibid., 166.

10. University of Virginia, Geospatial and Statistical Data Center, "Historical Census Browser," http://mapserver.lib.virginia.edu/.

11. University of Virginia, "Historical Census Browser." Data show that farms in the four-county area totaled 12,578 in 1910 and 10,432 in 1930, a 17 percent drop.

12. University of Virginia, "Historical Census Browser." These statistics are based on census data for Cherokee, Hall, Forsyth, and Jackson counties. U.S. Bureau of the Census, "Special Cotton Report," in *The Sixteenth Census of the United States, 1940* (Washington, D.C.: Government Printing Office, 1943).

13. Gilbert C. Fite, *Cotton Fields No More: Southern Agriculture, 1865–1980* (Lexington: University Press of Kentucky, 1984), 102.

14. U.S. Bureau of the Census, "Special Cotton Report"; Fite, *Cotton Fields No More*, 99, 87.

15. Arthur F. Raper, *Preface to Peasantry: A Tale of Two Black Belt Counties* (New York: Atheneum, 1936), 202–3. Rural sociologist Arthur F. Raper's *Preface to Peasantry* is a study of Greene and Macon counties in Georgia. For this study, Raper gathered data in these counties between 1927 and 1934. Greene and Macon counties lie south of the Upcountry. See also Range, *Century of Georgia Agriculture*, 174.

16. Range, *Century of Georgia Agriculture*, 174–75. On placing agricultural reform in the context of the Progressive Movement, see William A. Link, *The Paradox of Southern Progressivism, 1880–1930* (Chapel Hill: University of North Carolina Press, 1992); and George B. Tindall, *The Emergence of the New South, 1913–1945* (Baton Rouge: Louisiana State University Press, 1967).

17. Range, *Century of Georgia Agriculture*, 267–68, 175.

18. "Boll Weevil Coming to the Fair," *Gainesville News*, October 5, 1921.

19. On the boll weevil, see James C. Giesen, *Boll Weevil Blues: Cotton, Myth, and Power in the American South* (Chicago: University of Chicago Press, 2011).

20. "Down on Penny's Farm," on *We Won't Move: Songs of the Tenants' Movement* (Folkways Records & Service Corporation, 1983), housed in the Moses and Frances Asch Folkways Collection, Smithsonian Institution Center for Folklife and Cultural Heritage, Washington, D.C.

21. Andrew M. Soule, "Through Better Bread" (unpublished report, ca. 1920), Soule Papers, box 30.

22. Vance, *Human Factors in Cotton Culture*, 138; Range, *Century of Georgia Agriculture*; Charles S. Johnson, Edwin R. Embree, and W. W. Alexander, *The Collapse of Cotton Tenancy: Summary of Field Studies and Statistical Surveys, 1933–1935* (1935; repr., Chapel Hill: University of North Carolina Press, 2013), 47.

23. Range, *Century of Georgia Agriculture*, 272; Raper, *Preface to Peasantry*, 6, 77–78, 233, 256; Woodman, *King Cotton and His Retainers*; Harold Hoffsommer, "Survey of Rural Problem Areas: Morgan County, Georgia, Cotton Growing Region of the Old South" (Federal Emergency Relief Administration State Reports on Rural Problem Areas, 1934–35), iv–v, Records of the Division of Farm Population and Rural Life and Its Predecessors, Records of the Bureau of Agricultural Economics, RG 83, National Archives and Records Administration, College Park, Md. (hereafter NARA II); Arthur N. Moore, J. K. Giles, and R. C. Campbell, "Credit Problems of Georgia Cotton Farmers," Georgia Experiment Station Bulletin (Eatonton: Georgia Experiment Station of the University System of Georgia, 1929), 8.

24. Vance, *Human Factors in Cotton Culture*, 187, 189, 190. See also Rupert Bayless Vance, *Human Geography of the South: A Study in Regional Resources and Human Adequacy* (Chapel Hill: University of North Carolina Press, 1932). On the credit system, see Woodman, *King Cotton and His Retainers*, and Wright, *Old South, New South*.

25. Vance, *Human Factors in Cotton Culture*, 190.

26. University of Virginia, "Historical Census Browser."

27. Raper, *Preface to Peasantry*, 245; Vance, *Human Factors in Cotton Culture*, 190. On the plow-up, see also Pete Daniel, *Breaking the Land: The Transformation of Cotton, Tobacco, and Rice Cultures since 1880* (Urbana: University of Illinois Press, 1985), 258; Jack Temple Kirby, *Rural Worlds Lost: The American South, 1920–1960* (Baton Rouge: Louisiana State University Press, 1987), 58–65; Fite, *Cotton Fields No More*, 123, 129–30; Range, *Century of Georgia Agriculture*, 178.

28. Jimmy Carter, *An Hour before Daylight: Memories of a Rural Boyhood* (New York: Simon & Schuster, 2001), 70–71. Jimmy Carter's boyhood memoir chronicles his childhood in southwest Georgia. Carter did not come of age in the Upcountry, but the agricultural phenomenon on which he comments occurred throughout the South.

29. Charles McD. Puckette, "King Cotton's New Adventure: The South Watches with Hope Here and Misgiving There the Great Experiment in Which the First Step, Just Completed, Was the Plowing Under of Millions of Cultivated Acres," *New York Times Magazine*, August 27, 1933, sec. 6, 1.

30. Carter, *Hour before Daylight*, 64; Raper, *Preface to Peasantry*, 245.

31. U.S. Department of Agriculture, Agricultural Adjustment Administration, *Agricultural Adjustment: A Report of Administration of the Agricultural Adjustment Act May 1933 to February 1934* (Washington, D.C.: Government Printing Office, 1934); Theodore E. Whiting and Thomas Jackson Woofter, *Summary of Relief and Federal Work Program Statistics, 1933–1940* (Washington, D.C.: Government Printing Office, 1941); Miriam S. Farley, *Agricultural Adjustment under the New Deal* (New York: American Council, Institute of Pacific

Relations, 1936), 12–13; Fite, *Cotton Fields No More*, 132–33; Range, *Century of Georgia Agriculture*, 179; U.S. Bureau of the Census, "Special Cotton Report," xiv–xv.

32. T. L. Asbury, "Annual Report of the District 1 Agricultural Agent," 1935, 4–5, microfilm roll 71, Extension Service Annual Reports, Georgia, 1909–44, Annual Narrative and Statistical Reports, 1908–1974, Records of the Extension Service, RG 33, NARA II.

33. Raper, *Preface to Peasantry*, 245.

34. J. S. Stephenson, "Annual Narrative Report Extension, AAA, and Relief Activities in Hall County, December 1, 1935–December 1, 1936," 1936, 6, microfilm roll 82, Extension Service Annual Reports, Georgia, 1909–44, Annual Narrative and Statistical Reports, 1908–1974, Records of the Extension Service, RG 33, NARA II.

35. Johnson, Embree, and Alexander, *Collapse of Cotton Tenancy*, 51–53; Kirby, *Rural Worlds Lost*, 59; Raper, *Preface to Peasantry*, 245; Daniel, *Breaking the Land*, 105–8. The records of the U.S. Department of Agriculture's Extension Service confirm elite control. See, for example, Asbury, "Annual Report."

36. Johnson, Embree, and Alexander, *Collapse of Cotton Tenancy*, 52.

37. Guy Castleberry, interview by Lu Ann Jones, April 24, 1987, transcript, 22, Southern Agriculture Oral History Project, National Museum of American History, Washington, D.C. (hereafter SAOHP); H. A. Maxey, "Annual Report of Extension Activities, Cherokee County, December 31, 1934–December 31, 1935," 1935, 8, microfilm roll 72, Extension Service Annual Reports, Georgia, 1909–44, Annual Narrative and Statistical Reports, 1908–1974, Records of the Extension Service, RG 33, NARA II; H. A. Maxey, "Annual Narrative Report of Extension Activities, Cherokee County," 1937, 3 and 6–7, microfilm roll 89, Extension Service Annual Reports, Georgia, 1909–44, Annual Narrative and Statistical Reports, 1908–1974, Records of the Extension Service, RG 33, NARA II.

38. University of Virginia, "Historical Census Browser."

39. Range, *Century of Georgia Agriculture*, 274; Pete Daniel, "The Legal Basis of Agrarian Capitalism: The South since 1933," in *Race and Class in the American South since 1890*, ed. Melvyn Stokes and Rick Halpern (Providence, R.I.: Berg, 1994), 79–102; Raper, *Preface to Peasantry*, 252. Between 1930 and 1940, in nearby Cherokee County the number of tenant farms fell from 1,405 to 1,298. Over the same ten years, Forsyth County witnessed a similar decline as its tenant farms declined from 1,320 to 1,247. University of Virginia Library, Geospatial & Statistical Data Center, http://fisher.lib.virginia.edu/; U.S. Department of Agriculture, Agricultural Adjustment Administration, *Agricultural Adjustment*, 272.

40. Harold Hoffsommer, quoted in Johnson, Embree, and Alexander, *Collapse of Cotton Tenancy*, 58–59.

41. U.S. Bureau of the Census, "Special Cotton Report"; Tom Blackstock, interview by Lu Ann Jones, April 22, 1987, transcript, 10, 45, SAOHP.

42. John S. Jones and A. P. True, *A Search for Profits in Marketing Activities* (New York: Sales Management, 1938), 4–5.

43. U.S. Department of Agriculture, "Financing Production and Marketing of Broilers in the South: Part I, Dealer Phase," Southern Cooperative Series Bulletin no. 38 (June 1954), Agricultural Experiment Stations of Alabama, Arkansas, Georgia, Louisiana, Mississippi, North Carolina, South Carolina, Tennessee, Texas, and Virginia, and the Agricultural Marketing Service, 13.

44. W. W. Harper and O. C. Hester, "Influence of Production Practices on Marketing of Georgia Broilers," Georgia Agricultural Experiment Stations, University of Georgia College of Agriculture (1956), 9; W. W. Harper, "Marketing Georgia Broilers," Bulletin

no. 281 (July 1953), University of Georgia College of Agriculture Experiment Stations, 25; Gordon Sawyer, *The Agribusiness Poultry Industry: A History of Its Development* (New York: Exposition Press, 1971), 205.

45. Jones and True, *Search for Profits*, 4–5; Gordon Sawyer, of Gainesville, Ga., interview by author, July 15, 2003, tape recording, in author's possession; testimony of Jesse D. Jewell, May 13, 1957, Hearings before the Subcommittee No. 6 of the Select Committee on Small Business, House of Representatives, Eighty-Fifth Congress, First Session, 218–19; "An Interview with Jesse Jewell," *Broiler Industry*, March 1959, 3; "The Story of Jesse D. Jewell" (unpublished pamphlet, 1965), Jesse D. Jewell Vertical File, Georgia Mountain History Center, Gainesville.

46. Testimony of Jewell, 218–19.

47. Ted Oglesby, "Poultryland's Salute to Jesse Jewell," *Poultry Times*, November 7, 1971.

48. "Interview with Jesse Jewell."

49. Range, *Century of Georgia Agriculture*, 198–201; Sawyer, *Agribusiness Poultry Industry*, 85–95. The two main histories of the poultry industry are both written by industry insiders. Sawyer served as a public relations manager for poultry integrators in Hall County, Georgia. Oscar August Hanke was trained in poultry husbandry, wrote for the industry journals, served as president of the American Poultry Historical Society, and authored *American Poultry History, 1823–1973* (Lafayette, Ind.: American Poultry Historical Society, 1974).

50. Oglesby, "Poultryland's Salute"; "An Interview with Jesse Jewell," 3; Range, *Century of Georgia Agriculture*, 198–201; Sawyer, *Agribusiness Poultry Industry*, 85–95.

51. U.S. Department of Agriculture, "Financing Production," 49; Harper, "Marketing Georgia Broilers," 25; O. C. Hester and W. W. Harper, "The Function of Feed-Dealer Suppliers in Marketing Georgia Broilers," Bulletin no. 283 (August 1953), University of Georgia College of Agriculture Experiment Stations in cooperation with the Bureau of Agricultural Economics, U.S. Department of Agriculture, 11, 31; Harper and Hester, "Influence of Production Practices," 10.

52. James Aswell, "Chickens in the Wind," *Colliers*, September 9, 1950, 31–48; see also Sawyer, *Agribusiness Poultry Industry*, 85–96; Oglesby, "Poultryland's Salute."

53. Packers and Stockyards Administration, "The Broiler Industry: An Economic Study of Structure, Practices and Problems" (Washington, D.C.: U.S. Department of Agriculture, 1967), 1; Harper and Hester, "Influence of Production Practices," 5.

54. Harper and Hester, "Influence of Production Practices," 20; Verel W. Benson, Thomas J. Witzig, and Frederic Lewis Faber, "The Chicken Broiler Industry: Structure, Practices, and Costs," Agricultural Economic Report no. 381, Economic Research Service, U.S. Department of Agriculture (August 1977), 8; Hester and Harper, "Function of Feed-Dealer Suppliers," 11.

55. U.S. Department of Agriculture, "Financing Production," 13.

56. J. S. Stephenson, "Annual Narrative Report Extension," 1936, 4; J. W. Stephenson, "Annual Narrative Report Extension, AAA, and Relief Activities in Hall County," 1938, 7, microfilm roll 99, Extension Service Annual Reports, Georgia, 1909–44, Annual Narrative and Statistical Reports, 1908–1974, Records of the Extension Service, RG 33, NARA II.

57. Carter, *Hour before Daylight*, 89; Ruby Byers, interview by Lu Ann Jones, April 23, 1987, 9, SAOHP.

58. Spurgeon Welborn, interview by Lu Ann Jones, April 27, 1987, 11, 16, SAOHP; Carter, *Hour before Daylight*, 90; Ruby Byers, interview by Lu Ann Jones, 11.

59. "How I Developed My Market Project and What It Has Meant to Me by Mrs. O. H. Cooper, a Testimonial of a Club Woman, Walton county, Georgia," in Leila R. Mize, "Marketing Annual Report, 1937," 1937, unpaginated appendix, microfilm roll 87, Extension Service Annual Reports, Georgia, 1909–44, Annual Narrative and Statistical Reports, 1908–1974, Records of the Extension Service, RG 33, NARA II; "How I Developed My Market Project and What It Has Meant to Me by Mrs. W. D. Watson, Monroe County, Georgia," in Mize, "Marketing Annual Report," 1937, unpaginated appendix; Ray Marshall and Allen Thompson, "Status and Prospects of Small Farmers in the South," Southern Regional Council Report, Atlanta, Ga., 1976, 55; Harper and Hester, "Influence of Production Practices," 9; Blanche Whelchel, "Hall County, Home Dem. Agent Annual Report," 1933, 10, microfilm roll 60, Extension Service Annual Reports, Georgia, 1909–44, Annual Narrative and Statistical Reports, 1908–1974, Records of the Extension Service, RG 33, NARA II; testimonial of farm woman Mrs. B. H. Braden, quoted in Leila R. Mize, "Annual Report, Marketing Miscellaneous Home Products," microfilm roll 70, Extension Service Annual Reports, Georgia, 1909–44, Annual Narrative and Statistical Reports, 1908–1974, Records of the Extension Service, RG 33, NARA II.

60. Allie Street, "My Market Project and What It Has Meant to Me," in Mize, "Marketing Annual Report," 1937, unpaginated appendix.

61. Mrs. Mark Davis, "How I Developed My Market Project and What It Has Meant to Me by Mrs. Mark Davis, Floyd County, Georgia," in Mize, "Marketing Annual Report," 1937, unpaginated appendix.

62. Mrs. W. D. Watson, "How I Developed My Market Project," in Mize, "Marketing Annual Report," 1937, unpaginated appendix.

63. Gannon, "Early Fryer Project Outline," in "Poultry Husbandry Annual Report," 1937, unpaginated appendix, microfilm roll 87, Extension Service Annual Reports, Georgia, 1909–44, Annual Narrative and Statistical Reports, 1908–1974, Records of the Extension Service, RG 33, NARA II.

64. Arthur Flemming, interview by Lu Ann Jones, April 21, 1987, 4, 8, SAOHP; H. Y. Cook, "Annual Report of the Agricultural Extension Work, Hall County," 1935, 16, microfilm roll 74, Extension Service Annual Reports, Georgia, 1909–44, Annual Narrative and Statistical Reports, 1908–1974, Records of the Extension Service, RG 33, NARA II; Blackstock, interview by Lu Ann Jones, 11.

65. Sanford Byers, interview by Lu Ann Jones, April 20, 1987, 29, SAOHP.

66. "Interview with Jesse Jewell," 3; "Story of Jesse D. Jewell"; Ruby Byers, interview by Lu Ann Jones, 15.

67. Welborn, interview by Lu Ann Jones, 41–42; Ruby Byers, interview by Lu Ann Jones, 37; Lu Ann Jones, *Mama Learned Us to Work: Farm Women in the New South* (Chapel Hill: University of North Carolina Press, 2002), 104.

68. Welborn, interview by Lu Ann Jones, 13; Tom and Velva Blackstock, interview by Lu Ann Jones, April 22, 1987, SAOHP.

69. J. W. Jackson, "Jackson County, County Agent Annual Report," 1936, 7, NARA II, microfilm roll 83, Extension Service Annual Reports, Georgia, 1909–44, Annual Narrative and Statistical Reports, 1908–1974, Records of the Extension Service, RG 33, NARA II.

70. R. J. Richardson, "Poultry Marketing Annual Report—Project 7," 1938, 3–4, microfilm roll 96, Extension Service Annual Reports, Georgia, 1909–44, Annual Narrative and Statistical Reports, 1908–1974, Records of the Extension Service, RG 33, NARA II.

71. R. J. Richardson, "Poultry Marketing, Annual Report—Project 7," 1936, 5, microfilm roll 78, Extension Service Annual Reports, Georgia, 1909–44, Annual Narrative and Statistical Reports, 1908–1974, Records of the Extension Service, RG 33, NARA II.

72. Stephenson, "Annual Narrative Report Extension," 1937, 15, microfilm roll 91, Extension Service Annual Reports, Georgia, 1909–44, Annual Narrative and Statistical Reports, 1908–1974, Records of the Extension Service, RG 33, NARA II.

73. Gannon quoted in Stephenson, "Annual Narrative Report Extension," 1936, 6–7, 9.

74. Stephenson, "Annual Narrative Report Extension," 1936, 3; R. J. Richardson, "Poultry Marketing Annual Report," 1935, 7, microfilm roll 70, Extension Service Annual Reports, Georgia, 1909–44, Annual Narrative and Statistical Reports, 1908–1974, Records of the Extension Service, RG 33, NARA II.

75. Richardson, "Poultry Marketing Annual Report—Project 7," 1938, 4.

76. Richardson, "Poultry Marketing Annual Report—Project 7," 1935, 6.

77. J. C. Oglesbee, "Annual Report in Extension Agricultural Engineering," 1936, microfilm roll 78, Extension Service Annual Reports, Georgia, 1909–44, Annual Narrative and Statistical Reports, 1908–1974, Records of the Extension Service, RG 33, NARA II.

78. Oglesbee, "Annual Report in Extension Agricultural Engineering," 10; Oscar Steanson and Joe F. Davis, "Electricity on Farms in the Upper Piedmont of Georgia," Bulletin no. 263 (June 1950), Georgia Experiment Station of the University System of Georgia and the Bureau of Agricultural Economics.

79. Arthur Gannon, "Electricity as Related to Poultry" (pamphlet), in "Poultry Husbandry Annual Report," 1937, unpaginated appendix.

80. Captions of photos of E. L. Greeson's poultry farm, Whitfield County, Ga., J. L. Calhoun, "Rural Electrification Annual Report," 1938, unpaginated appendix, microfilm roll 96, Extension Service Annual Reports, Georgia, 1909–44, Annual Narrative and Statistical Reports, 1908–1974, Records of the Extension Service, RG 33, NARA II.

81. Lee C. Prickett, "Rural Electrification Annual Report," 1937, 4, microfilm roll 87, Extension Service Annual Reports, Georgia, 1909–44, Annual Narrative and Statistical Reports, 1908–1974, Records of the Extension Service, RG 33, NARA II; Ed Burch, "Farmers Using Electric Beds for Growing Potato Plants; Cheap Electricity Is a Boon," *Dalton Citizen*, June 10, 1937; "Practical Uses of Cheap Power Demonstrated by T. J. Cox at His Model Poultry Farm Near Dalton," *Dalton Citizen*, October 27, 1938, in Calhoun, "Rural Electrification Annual Report."

82. Captions of photos of E. L. Greeson's poultry farm, Whitfield County, Georgia, in Calhoun, "Rural Electrification Annual Report."

83. Gannon, "Electricity as Related to Poultry"; Prickett, "Rural Electrification Annual Report," 1937, 4.

84. "Practical Uses of Cheap Power."

85. Stephenson, "Annual Narrative Report Extension ," 1938, 16.

86. Peter L. Hansen and Ronald L. Mighell, "Economic Choices in the Broiler Industry" (Washington, D.C.: U.S. Department of Agriculture, 1956), 8.

87. William Boyd, "Making Meat: Science, Technology, and American Poultry Production," *Technology and Culture* 42, no. 4 (2001): 638–39.

88. Ibid., 641.

89. Arthur Gannon, Extension Poultryman, "Annual Report, Project 8-C—Poultry," December 1, 1935–November 30, 1936, 2, RG 33, micro copy T-855, roll 78, Extension Service Annual Reports/Georgia, 1909–44, 1935 (Wilcox Co.)–1936 (H. Dem. Ldr.).

90. Ibid., 3.

91. Arthur Gannon, Extension Poultryman, "Annual Report, Project 8-c—Poultry," December 1, 1934–November 30, 1935, RG 33, micro copy T-855, roll 70, Extension Service Annual Reports/Georgia, 1909–44, 1934 (Wayne Co.)–1935 (Director); Arthur Gannon, "Poultry Husbandry Annual Report," 1937, 2, RG 33.6, micro copy T-855, roll 87, Extension Service Annual Reports/Georgia, 1909–44, 1937 (Child Devel.—H. Demon. Dist. Agt.), NARA II; Arthur Gannon, "Poultry Husbandry Annual Report," 1940, appendix, 1, micro copy T-855, roll 111, Extension Service Annual Reports/Georgia, 1909–44, 1940 (Hortic.—H. Demon. Ldr.), RG 33.6, NARA II.

92. "Hall County Farm Statistics, 1900–1960," undated pamphlet from "Poultry Industry" vertical file, Hargrett Rare Books and Manuscripts, University of Georgia, Athens.

93. Arthur Gannon, "Annual Report Poultry Project No. 8-c for Calendar Year 1948," 1948, 2, Georgia Poultry, Annual Narrative and Statistical Reports, 1946–60, Records of the Extension Service, RG 33, NARA II; Gannon, "Annual Report Poultry Project No. 8-c for Calendar Year 1949," 1949, 2, Georgia Poultry, Annual Narrative and Statistical Reports, 1946–60, Records of the Extension Service, RG 33, NARA II.

94. H. A. Maxey, "Annual Narrative Report of Activities and Accomplishments in Cherokee County," 1950, 1, Annual Narrative and Statistical Reports, 1946–60, Records of the Extension Service, RG 33, NARA II.

95. Gannon, "Annual Report," 1947, 4; Arthur Gannon, "Annual Report Project No. 8-c Poultry," Georgia Poultry 1947, Annual Narrative and Statistical Reports, 1946–60, Records of the Extension Service, RG 33, NARA II, box 74.

96. Maxey, "Annual Narrative Report of Activities," 1.

97. Farley, *Agricultural Adjustment*, 21; William Alexander, foreword to Raper, *Preface to Peasantry,* xv.

98. R. J. Richardson, "Annual Report Project 7—Marketing Poultry, 1940," 47, micro copy T-855, roll 111, Extension Service Annual Reports/Georgia, 1909–44, 1940 (Hortic.—H. Demon. Ldr.), RG 33.6, NARA II.

99. John L. Anderson, "Annual Report of Extension Work Conducted in Jackson County, December 1, 1940 through November 30, 1941," 8, micro copy T-855, roll 121, Extension Service Annual Reports/Georgia, 1909–44, 1941 (Fayette–Johnson Cos.), RG 33.6, NARA II.

100. H. A. Maxey, "Annual Report of Extension Service Cherokee County, December 1, 1943 to December 1, 1944," 1944, 18, micro copy T-855, roll 137, Extension Service Annual Reports/Georgia, 1909–44, 1944 (Chatham–Grady Cos.), RG 33.6, NARA II.

101. John L. Anderson, "Jackson County Agricultural Agent Annual Report," 1945, 1–2, Jackson County, Ga., County Agent Annual Report, Annual Narrative and Statistical Reports, Records of the Extension Service, RG 33, NARA II.

102. John L. Anderson, "Annual Narrative Report of Activities and Accomplishments in Jackson County," 11, Jackson County, Ga., County Agent Annual Report 1946, Annual Narrative and Statistical Reports, 1946–60, Records of the Extension Service, RG 33, NARA II, box 77.

103. L. C. Rew, "Annual Narrative Report of Activities and Accomplishments in Hall County," 16, Hall County, Ga., County Agent Annual Report 1946, Annual Narrative and Statistical Reports, 1946–60, Records of the Extension Service, RG 33, NARA II.

104. Vance, *Human Factors in Cotton Culture*, 185.

105. J. H. Wood, "Poultry Possibilities of Georgia" (unpublished paper, 1927), Soule Papers, box 26.

Chapter 2. World War II and the Command Economy, 1939–1945

1. W. V. Chafin, "Goals for 'Food for Freedom' Drive Are Given to Farmers in Hall County by Wickard: Farmers Urged to Insure Full Dinner Pail for All—Americans and Allies," in W. V. Chafin, "Annual Narrative Report of Agricultural Extension Work in Hall County, December 1, 1940 to November 30, 1941," 1941, 11, microfilm roll 121, Extension Service Annual Reports, Georgia, 1909–44, Annual Narrative and Statistical Reports, 1908–1974, Records of the Extension Service, RG 33, NARA II.

2. W. W. Harper, "Marketing Georgia Broilers," Bulletin no. 281 (July 1953), University of Georgia College of Agriculture Experiment Stations, 23.

3. Ibid., 26–27.

4. "Hall County Farm Statistics, 1900–1960," undated pamphlet from "Poultry Industry" vertical file, Hargrett Rare Books and Manuscripts, University of Georgia, Athens.

5. H. A. Maxey, "Annual Narrative Report of Activities and Accomplishments in Cherokee County," 1950, 1, Annual Narrative and Statistical Reports, 1946–60, Records of the Extension Service, RG 33, NARA II.

6. Arthur Gannon, "Poultry Husbandry Annual Report," 1940, 13, microfilm roll 111, Extension Service Annual Reports, Georgia, 1909–44, Annual Narrative and Statistical Reports, 1908–1974, Records of the Extension Service, RG 33, NARA II; Arthur Gannon, "Annual Report Poultry Project No. 8-C," 1942, 10, microfilm roll 127, Extension Service Annual Reports, Georgia, 1909–44, Annual Narrative and Statistical Reports, 1908–1974, Records of the Extension Service, RG 33, NARA II.

7. Catherine V. Wood, "Annual Report Poultry Project No. 8-C," 1943, 9, microfilm roll 141, Extension Service Annual Reports, Georgia, 1909–44, Annual Narrative and Statistical Reports, 1908–1974, Records of the Extension Service, RG 33, NARA II.

8. B. T. Brown and B. H. Kinney, "Forsyth County Narrative Report," 1940, 11, microfilm roll 115, Extension Service Annual Reports, Georgia, 1909–44, Annual Narrative and Statistical Reports, 1908–1974, Records of the Extension Service, RG 33, NARA II.

9. B. T. Brown, "Forsyth County Narrative Report," 1941, 2, 13, microfilm roll 121, Extension Service Annual Reports, Georgia, 1909–44, Annual Narrative and Statistical Reports, 1908–1974, Records of the Extension Service, RG 33, NARA II.

10. H. A. Maxey, "Annual Narrative Report of Extension Service, Cherokee County, December 1, 1940 to December 1, 1941," 1941, 10, micro copy T-855, roll 137, Extension Service Annual Reports/Georgia, 1909–44, 1941 (Bibb & Twiggs–Clay Co.), RG 33, NARA II.

11. Arthur Gannon, "Annual Report, Project 8-C–Poultry," 1941, 11, Georgia Agricultural Extension Service, micro copy T-855, roll 120, Extension Service Annual Reports/Georgia, 1909–44, 1941 (Landscaping–Berrien Co.), RG 33, NARA II.

12. Wood, "Annual Report Poultry Project No. 8-C," 1943, 9.

13. Chafin, "Annual Narrative Report," 16, and unpaginated appendix.

14. L. S. Watson, "Annual Report, Northwest Georgia," 1940, 1, microfilm roll 111, Extension Service Annual Reports, Georgia, 1909–44, Annual Narrative and Statistical Reports, 1908–1974, Records of the Extension Service, RG 33, NARA II; Chafin, "Annual Narrative Report," 1941, 5.

15. Chafin, "Annual Narrative Report," 1941, unpaginated appendix; "Georgia Agricultural Outlook for 1941," quoted in Leila R. Mize, "Home Marketing Annual Report," 1940, 2–3, microfilm roll III, Extension Service Annual Reports, Georgia, 1909–44, Annual Narrative and Statistical Reports, 1908–1974, Records of the Extension Service, RG 33, NARA II.

16. "Hall County Farm Statistics, 1900–1960."

17. H. A. Maxey, "Annual Narrative Report of Activities and Accomplishments in Cherokee County," 1947, 12, Annual Narrative and Statistical Reports, 1946–60, Records of the Extension Service, RG 33, NARA II.

18. Susan Mathews, "Annual Report, Nutrition Project," 1941, 1, microfilm roll 119, Extension Service Annual Reports, Georgia, 1909–44, Annual Narrative and Statistical Reports, 1908–1974, Records of the Extension Service, RG 33, NARA II.

19. Interbureau Committee on Post-War Programs, "Farmers Look at Post-War Prospects," 1945, 17, Farmers Appraisal of Post-war Problems Study no. 85, Records of the Division of Program Surveys, Project Files, 1940–45, Records of the Bureau of Agricultural Economics, RG 83, NARA II.

20. Susan Matthews, "Annual Report Nutrition Project," 1940, 8, microfilm roll 110, Extension Service Annual Reports, Georgia, 1909–44, Annual Narrative and Statistical Reports, 1908–1974, Records of the Extension Service, RG 33, NARA II; W. A. Hartman, "Estimate of Need of Medical Care in Rural Areas," 1941, 3 and 10, Post-War Planning, Rural Health 1941–1942, Southeast Regional Office, Atlanta, Georgia, Records of the Post-War Planning Commission, 1941–45, Objectives and Assumptions, 1942–43 to Wartime Production Adjustments in the South, Records of the Bureau of Agricultural Economics, RG 83, National Archives Southeast Region, Morrow, Ga.

21. Matthews, "Annual Report Nutrition Project," 1941, 1.

22. H. A. Maxey, "Annual Report of Extension Service Cherokee County, December 1, 1943 to December 1, 1944," 1944, 22, microfilm roll 137, Extension Service Annual Reports, Georgia, 1909–44, Annual Narrative and Statistical Reports, 1908–1974, Records of the Extension Service, RG 33, NARA II; John L. Anderson, "Narrative Annual Report of Extension Work Conducted in Jackson County, Georgia," 1944, 10, microfilm roll 138, Extension Service Annual Reports, Georgia, 1909–44, Annual Narrative and Statistical Reports, 1908–1974, Records of the Extension Service, RG 33, NARA II.

23. Martha McAlpine, "Child Development and Family Life Annual Report," 1943, 2, microfilm roll 131, Extension Service Annual Reports, Georgia, 1909–44, Annual Narrative and Statistical Reports, 1908–1974, Records of the Extension Service, RG 33, NARA II; Numan V. Bartley, *The Creation of Modern Georgia* (Athens: University of Georgia Press, 1983), 180.

24. Velva Blackstock, interview by Lu Ann Jones, April 22, 1987, 42, Southern Agriculture Oral History Project, National Museum of American History, Washington, D.C. (hereafter SAOHP).

25. Interbureau Committee on Post-War Programs, "Farmers Look at Post-War Prospects," 1945, 3.

26. Kenneth Treanor and J. W. Fanning, "Annual Report, Farm Management Extension Work, Project no. 7," 1942, 16, microfilm roll 126, Extension Service Annual Reports, Georgia, 1909–44, Annual Narrative and Statistical Reports, 1908–1974, Records of the Extension Service, RG 33, NARA II.

27. Lula Edwards, "Annual Report Northwest Georgia District" 1942, 7, microfilm roll 127, Extension Service Annual Reports, Georgia, 1909–44, Annual Narrative and Statistical Reports, 1908–1974, Records of the Extension Service, RG 33, NARA II.

28. Pete Daniel, *Breaking the Land: The Transformation of Cotton, Tobacco, and Rice Cultures since 1880* (Urbana: University of Illinois Press, 1985), 243.

29. H. A. Maxey, "Annual Narrative Report of Extension Service, Cherokee County, December 1, 1944 to December 1, 1945," 1945, 22, Cherokee County, Georgia, County Agent Annual Report, Annual Narrative and Statistical Reports 1945, Records of the Extension Service, RG 33, NARA II.

30. L. C. Rew, "Narrative Annual Report of Extension Work Conducted in Hall County, Georgia," 1944, 19, microfilm roll 138, Extension Service Annual Reports, Georgia, 1909–44, Annual Narrative and Statistical Reports, 1908–1974, Records of the Extension Service, RG 33, NARA II.

31. Anderson, "Annual Report," 1944, 12.

32. "Migration Scouting—Floyd County, Georgia," 1944, 3, unpublished report, Records of the Division of Program Surveys, Project Files, 1940–45, Records of the Bureau of Agricultural Economics, RG 83, NARA II. The name of the farmer interviewed was not given.

33. L. S. Watson, "Annual Report Northwest Georgia for Calendar Year 1945," 1945, 20–21, Georgia County Agent Leader Annual Report, Annual Narrative and Statistical Reports, Records of the Extension Service, RG 33, NARA II.

34. L. C. Rew and W. V. Chafin, "Annual Report Hall County, Georgia," 1943, 5, microfilm roll 133, Extension Service Annual Reports, Georgia, 1909–44, Annual Narrative and Statistical Reports, 1908–1974, Records of the Extension Service, RG 33, NARA II; John L. Anderson, "Jackson County Agricultural Agent Annual Report," 1945, 7, Jackson County, Ga., County Agent Annual Report, Annual Narrative and Statistical Reports, Records of the Extension Service, RG 33, NARA II. Women entered the fields as well. See Ida Bell, "Annual Report Northwest Georgia District," 1944, 1, microfilm roll 136, Extension Service Annual Reports, Georgia, 1909–44, Annual Narrative and Statistical Reports, 1908–1974, Records of the Extension Service, RG 33, NARA II.

35. Anderson, "Annual Report," 1945, 4.

36. Bureau of Agricultural Economics, U.S. Department of Agriculture, "Impressions Regarding Negroes," September 27, 1941, "Special Reports re Negroes," Records of the Division of Program Surveys, Project Files, 1940–45, Records of the Bureau of Agricultural Economics, RG 83, NARA II.

37. See Cindy Hahamovitch, *The Fruits of Their Labor: Atlantic Coast Farmworkers and the Making of Migrant Poverty, 1870–1945* (Chapel Hill: University of North Carolina Press, 1997); on reduction in the number of farm laborers, see Kenneth Treanor and J. W. Fanning, "Annual Report Farm Management Extension Work Project No. 7 for Calendar Year 1943," 1943, 3, microfilm roll 131, Extension Service Annual Reports, Georgia, 1909–44, Annual Narrative and Statistical Reports, 1908–1974, Records of the Extension Service, RG 33, NARA II.

38. Treanor and Fanning, "Annual Report," 1943, 3.

39. Maxey, "Annual Narrative Report," 1945, 20–21.

40. Claude R. Wickard, *Report of the Secretary of Agriculture* (Washington, D.C.: Government Printing Office, U.S. Department of Agriculture, 1941), 157; Kenneth Treanor, Charles E. Clark, W. A. King, and Harry A. White, "Annual Report Farm Management

Extension Work," 1946, 13, Georgia, Farm Management 1946, Annual Narrative and Statistical Reports, 1946–60, Records of the Extension Service, RG 33, NARA II.

41. Wickard, *Report of the Secretary of Agriculture*, 1941, 157.

42. Treanor and Fanning, "Annual Report," 1942, 5.

43. Pete Daniel, "A Hundred Years of Dispossession: Southern Farmers and the Forces of Change," in *Outstanding in His Field: Perspectives on American Agriculture in Honor of Wayne D. Rasmussen*, ed. Frederick Carstensen, Morton Rothstein, and Joseph Swanson (Ames: Iowa State University Press, 1993), 90, 96.

44. Anderson, "Annual Report," 1944, 10; see also on the rise of mechanization, Watson, "Annual Report," 1945, 12; Anderson, "Annual Report," 1945, 20; L. C. Rew, "Annual Narrative Report of Agriculture Extension Work in Hall County, Georgia, December 1, 1944 to December 1, 1945," 1945, 13, Hall County, Georgia, County Agent Annual Report 1945, Annual Narrative and Statistical Reports, Records of the Extension Service, RG 33, NARA II.

45. Daniel, *Breaking the Land*, 6.

46. Spurgeon Welborn, interview by Lu Ann Jones, April 27, 1987, 8, SAOHP.

47. Ibid., 47.

48. Sanford Byers, interview by Lu Ann Jones, April 20, 1987, 17, SAOHP.

49. H. A. Maxey, "Annual Report of Extension Service Cherokee County, December 1, 1941 to December 1, 1942," 1942, 2, micro copy T-855, roll 120, Extension Service Annual Reports/Georgia, 1909–44, 1942 (Catoosa–Gilmer Cos.), RG 33.6, NARA II; H. A. Maxey, "Annual Report of Extension Service Cherokee County, December 1, 1942 to December 1, 1943," 1943, 15, micro copy T-855, roll 132, Extension Service Annual Reports/Georgia, 1909–44, 1943 (Baldwin–Decatur Cos.), RG 33.6, NARA II; H. A. Maxey, "Annual Report of Extension Service Cherokee County, December 1, 1943 to December 1, 1944," 1944, 19, micro copy T-855, roll 137, Extension Service Annual Reports/Georgia, 1909–44, 1943 (Chatham–Grady Cos.), RG 33.6, NARA II; John L. Anderson, "Annual Report of Extension Work Conducted in Jackson County, December 1, 1940 through November 30, 1941," 1941, 44, micro copy T-855, roll 121, Extension Service Annual Reports/Georgia, 1909–44, 1941 (Fayette–Johnson Cos.), RG 33.6, NARA II; John. L. Anderson, "Annual Report of Extension Work in Jackson County, Georgia," 1942, 24–25, micro copy T-855, roll 129, Extension Service Annual Reports/Georgia, 1909–44, 1943 (Glascock–Mitchell Cos.), RG 33.6, NARA II.

50. "Informational Summary Georgia Agricultural Extension Service Week Ending October 24, 1942," appendix A in O. B. Copeland, "Annual Report Publicity Project No. 2," 1942, 6 of Appendix A, microfilm roll 127, Extension Service Annual Reports, Georgia, 1909–44, Annual Narrative and Statistical Reports, 1908–1974, Records of the Extension Service, RG 33, NARA II.

51. Claude R. Wickard, *Report of the Secretary of Agriculture* (Washington, D.C.: Government Printing Office, U.S. Department of Agriculture, 1944), 135.

52. Gannon, "Annual Report," 1940, no. 112, 15.

53. R. J. Richardson, "Annual Report, Project 7, Poultry Marketing," 1941, 21, micro copy T-855, roll 120, Extension Service Annual Reports/Georgia, 1909–44, 1941 (Landscaping–Berrien Co.), RG 33.6, NARA II; Camilla Weems, "Annual Narrative Report, Negro Home Demonstration Work," 1941, 81, micro copy T-855, roll 120, Extension Service Annual Reports/Georgia, 1909–44, 1941 (Landscaping–Berrien Co.), RG 33.6, NARA II.

54. Grover B. Hill to Homer M. Adkins, February 17, 1942, Agricultural Adjustment Program 4-2-15, Poultry—Poultry Products, January 1–March 27, 1942, General Corre-

spondence of the Office of the Secretary, 1906–70, Records of the Office of the Secretary of Agriculture, RG 16, NARA II.

55. Gannon, "Annual Report," 1941; Grover B. Hill to O. H. Hanke, August 17, 1943, Correspondence of the War Food Administration, 1943–45, Records of the War Food Administration, Records of the Office of the Secretary of Agriculture, NARA II; "What Can Georgia Farm Families Do to Make Their Best Contribution to the War Effort?," pamphlet included in appendix to Gannon, "Annual Report," 1942; Hill to Adkins, February 17, 1942.

56. Grover B. Hill to Lindley Beckworth, April 1, 1944, Correspondence of the War Food Administration, 1943–45, Records of the War Food Administration, Records of the Office of the Secretary of Agriculture, RG 16, NARA II, 1 of 2; Emery E. Jacobs to O. R. Jones, February 21, 1942, Agricultural Adjustment Program 4-2-15, Poultry—Poultry Products, Jan 1–March 27, 1942, General Correspondence of the Office of the Secretary, 1906–70 (entry 17), Records of the Office of the Secretary of Agriculture, RG 16, NARA II, box 597; Bureau of Animal Industry, Agricultural Research Administration, "Report on Defense and War Activities of the Bureau of Animal Industry, for Period January 1–March 31, 1943," April 1943, 4, Bureau of Animal Industry, Quarterly Reports for the War Records Project, 1941–45, Records of the Division of Statistical and Historical Research, Records of the Bureau of Agricultural Economics, RG 83, NARA II.

57. Maxey, "Annual Report," 1943, 6; Walter W. Wilcox, *The Farmer in the Second World War* (Ames: Iowa State College Press, 1947); Grover B. Hill to Bennet Champ Clark, U.S. senator from Arkansas, March 26, 1942, Agricultural Adjustment Program 4-2-15, Poultry—Poultry Products, January 1–March 27, 1942, General Correspondence of the Office of the Secretary, 1906–70, Records of the Office of the Secretary of Agriculture, RG 16, NARA II.

58. Hill to Hanke, August 17, 1943.

59. "What Can Georgia Farm Families Do?"; Jacobs to Jones, February 21, 1942; Grover B. Hill to Richard B. Russell, April 3, 1942, Agricultural Adjustment Program 4-2-15, Poultry—Poultry Products, January 1–March 27, 1942, General Correspondence of the Office of the Secretary, 1906–70, Records of the Office of the Secretary of Agriculture, RG 16, NARA II.

60. Letter pasted into Gannon, "Annual Report," 1941.

61. Grover B. Hill to Stephen Pace, House of Representatives, March 11, 1942, Agricultural Adjustment Program 4-2-15, Poultry—Poultry Products, January 1–March 27, 1942, General Correspondence of the Office of the Secretary, 1906–70, Records of the Office of the Secretary of Agriculture, RG 16, NARA II; Paul H. Appleby to Lyndon B. Johnson, March 22, 1943, Agricultural Adjustment Program 4-2-15, Poultry—Poultry Products, General Correspondence of the Office of the Secretary, 1906–70, Records of the Office of the Secretary of Agriculture, RG 16, NARA II; Hill to Clark, March 26, 1942.

62. Hill to Russell, April 3, 1942.

63. Grover B. Hill to Hattie W. Caraway, U.S. senator from Arkansas, March 20, 1942, Agricultural Adjustment Program 4-2-15, Poultry—Poultry Products, January 1–March 27, 1942, General Correspondence of the Office of the Secretary, 1906–70, Records of the Office of the Secretary of Agriculture, RG 16, NARA II; Grover B. Hill to David D. Terry, U.S. representative from Arkansas, February 16, 1942, Agricultural Adjustment Program 4-2-15, Poultry—Poultry Products, January 1–March 27, 1942, General Correspondence of the Office of the Secretary, 1906–70, Records of the Office of the Secretary of Agriculture, RG 16, NARA II.

64. Arthur F. Peine and Ernest O. Eisenberg to James P. Calvin, March 11, 1943, Poultry, January 1–April 1, 1943, General Correspondence of the Office of the Secretary, 1906–70, Records of the Office of the Secretary of Agriculture, RG 16, NARA II.

65. "Informational Summary Georgia Agricultural Extension Service," 7 of appendix A; Claude R. Wickard to O. A. Hanke, editor of the *Poultry Tribune*, September 15, 1942, Agricultural Adjustment Program 4-2-15, Poultry—Poultry Products, January 1–March 27, General Correspondence of the Office of the Secretary, 1906–70, Records of the Office of the Secretary of Agriculture, RG 16, NARA II.

66. Claude R. Wickard, *Report of the Secretary of Agriculture* (Washington, D.C.: Government Printing Office, U.S. Department of Agriculture, 1943), 155; "Title 32—National Defense Chapter XI—Office of Price Administrator Part 1429. Poultry and Eggs Revised Maximum Price Regulation No. 269 Poultry," 1942, Poultry, January–December 1942, General Correspondence of the Office of the Secretary, 1906–70, Records of the Office of the Secretary of Agriculture, RG 16, NARA II.

67. Rew and Chafin, "Annual Report Hall County, Georgia," 8.

68. Arthur Gannon, "Report for January, February, March 1943—Project 8-C Poultry," 1943, 15, T-855, roll 131, Extension Service Annual Reports/Georgia, 1909–44, 1942 (Twiggs Co.)–1943 (Baker Co.), RG 33.6, NARA II; D. L. Floyd, "Comments to Accompany Report of Agricultural Statistician, Poultry—June Rural Carrier Livestock Survey, Georgia," 1943, States Comments, June Rural Carrier, 1938–44, Divisions Responsible for Agricultural Estimates, Records Relating to Poultry Surveys, 1931–1949, Records of the Bureau of Agricultural Economics, RG 83, NARA II.

69. Wickard, *Report of the Secretary of Agriculture*, 1943, 155.

70. Wickard, *Report of the Secretary of Agriculture*, 1944, 135; Lee Marshall to T. G. Stitts and L. T. Hopkinson, December 9, 1944, Poultry, from October 1944 to December 1944, Correspondence of the War Food Administration, 1943–45, Records of the War Food Administration, Records of the Office of the Secretary of Agriculture, RG 16, NARA II; Grace N. Adams, "Hall County Home Demonstration Agent Annual Report 1945," 1945, 5, Hall County, Ga., Home Demonstration Agent Annual Report, Annual Narrative and Statistical Reports, Records of the Extension Service, RG 33, NARA II.

71. Archie Langley, "Georgia 1944 Commercial Broiler Production," 1949, 3, States' Summaries and Comments, Commercial Broiler Production, 1940–44, Divisions Responsible for Agricultural Estimates, Records Relating to Poultry Surveys, 1931–1949, Records of the Bureau of Agricultural Economics, RG 83, NARA II.

72. Peine and Eisenberg to Calvin, March 11, 1943, 4; Floyd, "Comments to Accompany Report of Agricultural Statistician"; Rew and Chafin, "Annual Report Hall County, Georgia," 7; Roger Horowitz, "Making the Chicken of Tomorrow: Reworking Poultry as Commodities and as Creatures, 1945–1990," in *Industrializing Organisms: Introducing Evolutionary History*, ed. Philip and Susan R. Scranton (New York: Routledge, 2003), 215–35, 218.

73. Marvin P. Jones to Fred M. Vinson, May 17, 1945, Poultry, from May 1, 1945, to June 30, 1945, Correspondence of the War Food Administration, 1943–45, Records of the War Food Administration, Records of the Office of the Secretary of Agriculture, RG 16, NARA II.

74. Ashley Sellers to J. Monroe Johnson, May 14, 1945, Poultry, from May 1, 1945, to June 30, 1954, Correspondence of the War Food Administration, 1943–45, Records of the War Food Administration, Records of the Office of the Secretary of Agriculture, RG 16, NARA II; Grover B. Hill to J. Monroe Johnson, December 29, 1944, Poultry, from October 1944 to December 1944, Correspondence of the War Food Administration, 1943–45,

Records of the War Food Administration, Records of the Office of the Secretary of Agriculture, RG 16, NARA II.

75. Clinton P. Anderson to Arthur Capper, August 17, 1945, Poultry (2 of 3), August 16–30, 1945, General Correspondence of the Office of the Secretary, 1906–70, Records of the Office of the Secretary of Agriculture, RG 16, NARA II.

76. Jones to Vinson, May 17, 1945.

77. Brooklyn Poultry Dealers Association to Clinton P. Anderson, Poultry (1 of 3), January 1–August 15, 1945, General Correspondence of the Office of the Secretary, 1906–70, Records of the Office of the Secretary of Agriculture, RG 16, NARA II.

78. Horowitz, "Making the Chicken of Tomorrow," 218; Wilcox, *Farmer in the Second World War*, 154.

79. Arthur Gannon, "Poultry Husbandry Annual Report," 1937, 2, microfilm roll 87, Extension Service Annual Reports, Georgia, 1909–44, Annual Narrative and Statistical Reports, 1908–1974, Records of the Extension Service, RG 33, NARA II.

80. Gannon, "Annual Report," 1940, unpaginated appendix, 15; Arthur Gannon, "Annual Report Poultry Project No. 8-C for Calendar Year 1945," 1945, 3, Georgia Poultry 1945, Annual Narrative and Statistical Reports, 1945 (entry 11), RG 33, NARA II; Arthur Gannon, "Annual Report Poultry Project No. 8-C for Calendar Year 1946," 1946, 4, Georgia Poultry 1946, Annual Narrative and Statistical Reports, 1946–60, RG 33, NARA II.

81. Bureau of Animal Industry, "Report on Defense and War Activities of the Bureau of Animal Industry, Agricultural Research Administration, USDA for Period July 1–September 30, 1942," 1942, 6, Agricultural Adjustment Administration—Bureau of Dairy Industry, Quarterly Reports for the War Records Project, 1941–45, Records of the Division of Statistical and Historical Research, Records of the Bureau of Animal Industry, Records of the Bureau of Agricultural Economics, RG 83.

82. Rew, "Annual Report," 1945, 4–5; John S. Wood to Clinton P. Anderson, August 7, 1945, Poultry (2 of 3) August 16–30, 1945, General Correspondence of the Office of the Secretary, 1906–70 (entry 17), RG 16, NARA II.

83. R. J. Richardson, "Annual Report, Project 7, Poultry Marketing," 1941, microfilm roll 120, Extension Service Annual Reports, Georgia, 1909–44, Annual Narrative and Statistical Reports, 1908–1974, Records of the Extension Service, RG 33, NARA II.

84. R. J. Richardson, "Annual Report, Project 7 (Marketing Poultry)," 1942, 24, microfilm roll 127, Extension Service Annual Reports, Georgia, 1909–44, Annual Narrative and Statistical Reports, 1908–1974, Records of the Extension Service, RG 33, NARA II.

85. B. T. Brown and W. L. Hawes, "Forsyth County Narrative Report," 1943, 3, microfilm roll 132, Extension Service Annual Reports, Georgia, 1909–44, Annual Narrative and Statistical Reports, 1908–1974, Records of the Extension Service, RG 33, NARA II.

86. Rew, "Annual Report," 1944, 7; Rew, "Annual Report," 1945, 9, Annual Narrative and Statistical Reports, 1946–60, Records of the Extension Service, RG 33, NARA II; "Hall County, Georgia, County Agent Annual Report," 1946, 5, Annual Narrative and Statistical Reports, 1946–60, Records of the Extension Service, RG 33, NARA II.

87. "Evisceration Methods in Dressing Explored," undated newspaper clipping in L. C. Rew, "Annual Narrative Report of Activities and Accomplishments in Hall County," 1947, Hall County Georgia County Agent Annual Report, 1947, Annual Narrative and Statistical Reports, 1946–60, Records of the Extension Service, RG 33, NARA II.

88. Frederick Produce to Thomas Albert, February 22, 1945, Poultry, from January 1, 1945, to April 30, 1954, Correspondence of the War Food Administration, 1943–45, Records

of the War Food Administration, Records of the Office of the Secretary of Agriculture, RG 16, NARA II.

89. Rew, "Annual Report," 1943, 9; Rew, "Annual Report," 1944, 8.

90. Anderson, "Annual Report," 1945, 1.

91. Ward Keller to John Cochron, April 21, 1945, Poultry, from May 1, 1945, to June 30, 1954, Correspondence of the War Food Administration, 1943–45, Records of the War Food Administration, Records of the Office of the Secretary of Agriculture, RG 16, NARA II.

92. Ibid.

93. Rew, "Annual Report," 1943, 9.

94. L. C. Rew, Annual Narrative Report of Activities and Accomplishments in Hall County," 1946, 5, Hall County Georgia County Agent Annual Report, 1946, Annual Narrative and Statistical Reports, 1946–60, Records of the Extension Service, RG 33, NARA II.

95. Ibid., 3.

96. "Clower Says Poultry Lab Opens Soon: Extension New Specialist Praises New Building and Modern Facilities," undated newspaper clipping pasted into L. C. Rew, "Annual Narrative Report of Activities and Accomplishments in Hall County," 1947, Hall County Georgia County Agent Annual Report 1947, Annual Narrative and Statistical Reports, 1946–60, RG 33, NARA II; Maxey, "Annual Narrative Report," 1945, 12; John L. Anderson, "Annual Narrative Report of Activities and Accomplishments in Jackson County," 1946, 8, Jackson County, Georgia, County Agent Annual Report 1946, Annual Narrative and Statistical Reports, 1946–60, RG 33, NARA II; H. A. Maxey, "Annual Report of Extension Service Cherokee County, December 1, 1941 to December 1, 1942," 1942, 10, micro copy T-855, roll 128, Extension Service Annual Reports/Georgia, 1909–44, 1942 (Catoosa–Gilmer Cos.), RG 33.6, NARA II.

97. R. J. Richardson, "Annual Report, Project 7, Poultry Marketing," 1941, microfilm roll 120, Extension Service Annual Reports, Georgia, 1909–44, Annual Narrative and Statistical Reports, 1908–1974, Records of the Extension Service, RG 33, NARA II.

98. Camilla Weems, "Annual Narrative Report, Negro Home Demonstration Work," 1941, 6, micro copy T-855, roll 120, Extension Service Annual Reports/Georgia, 1909–44, 1941 (Landscaping–Berrien Co.), RG 33.6, NARA II.

99. Ibid., 3.

100. Ibid., 8.

101. Arthur Gannon, "Annual Report Poultry Project No. 8-C for Calendar Year 1948," 1948, 9, Georgia Poultry, Annual Narrative and Statistical Reports, 1946–60, Records of the Extension Service, RG 33, NARA II.

102. Claude R. Wickard, "Statement of the Secretary of Agriculture Claude R. Wickard at a Hearing of the House Special Committee on Post-War Economic Policy and Planning Opening at 10:00am, August 23, 1944," EN 516, p. 6, Post-War Planning, Reports of Working Committee, 1944–45, Southeast Regional Office, Atlanta, Ga., Records of the Post-War Planning Commission, 1941–45, Objectives and Assumptions, 1942–43 to Wartime Production Adjustments in the South, 1942, RG 83, Bureau of Agricultural Economics Repository, National Archives Southeast Region, Morrow, Ga.

103. Treanor and Fanning, "Annual Report," 1942, 16. See Kari Frederickson's article "Confronting the Garrison State: South Carolina in the Early Cold War Era," *Journal of Southern History* 72, no. 2 (2006): 349–78.

104. Bureau of Agricultural Economics, U.S. Department of Agriculture, "Report

on Disposition of Land Temporarily Used by Military Forces and War Plants," 1943, 1, Conference Committee no. VI, Milwaukee Conference on Post-War Programs, Report on Industry in Rural Areas, Handbook of Post-War Planning Materials, 1941–1945 (file 2 of 2), Records of the Southeastern Regional Office, Atlanta, Ga., Records of the Post-War Planning Commission, 1941–45, Handbook of Post-War Planning Materials, 1941–1945 to Correspondence 1945, RG 83, Bureau of Agricultural Economics, National Archives Southeast Region, Morrow, Ga.

105. Bureau of Agricultural Economics, U.S. Department of Agriculture, "The Philosophy of Adjustment for Southeastern Agriculture," 1943, 14, Correspondence Postwar Planning Committee, 1943, Southeast Regional Office, Atlanta, Ga., Subgroup: Records of the Post-War Planning Commission, 1941–45, Correspondence, 1943 to Regional Committees, 1942–43, RG 83, Bureau of Agricultural Economics, National Archives Southeast Region, Morrow, Ga.

106. Ibid., 13.

107. Ibid., 16.

108. Daniel, "Hundred Years of Dispossession," 90, 96. "During the war these advocates of a mechanized, highly commercialized agriculture helped initiate an abrupt two-decade shift to machines and wage labor. Superseding the labor-intensive sharecropping that had defined southern agriculture since Reconstruction, the new farm system was the fruit of a revolution." Pete Daniel, "Going among Strangers: Southern Reactions to World War II," *Journal of American History* 77, no. 3 (1990): 886–911, 888.

109. Rew, "Narrative Annual Report," 1944, 19.

110. Bureau of Agricultural Economics, U.S. Department of Agriculture, "Suggested List of Subjects for the Department's Program of Work on Post War Problems in the Present Fiscal Year," 1943, 3, Milwaukee Conference on Post-War Programs, July 26–31; Report on Credit by Committee no. VI, Handbook of Post-War Planning Materials, 1941–1945 (file 2 of 2), Records of the Southeastern Regional Office, Atlanta, Ga., Records of the Post-War Planning Commission, 1941–45, Handbook of Post-War Planning Materials, 1941–1945 to Correspondence 1945, RG 83, Bureau of Agricultural Economics, National Archives Southeast Region, Morrow, Ga.

111. Frederickson, "Confronting the Garrison State," 355.

112. Ibid., 353.

113. Ibid.; Morton Sosna, "More Important Than the Civil War? The Impact of World War II on the South," *Perspectives on the American South* 4 (1987): 145–61; Russell B. Olwell, *At Work in the Atomic City: A Labor and Social History of Oak Ridge, Tennessee* (Knoxville: University of Tennessee Press, 2004); James C. Cobb, *The Selling of the South: The Southern Crusade for Industrial Development, 1936–1990* (Champaign: University of Illinois Press, 1993); Bruce J. Schulman, *From Cotton Belt to Sunbelt: Federal Policy, Economic Development, and the Transformation of the South, 1938–1980* (Durham, N.C.: Duke University Press, 1994); Philip Scranton, ed., *The Second Wave: Southern Industrialization from the 1940s to the 1970s* (Athens: University of Georgia Press, 2001); William Barnaby Faherty, *Florida's Space Coast: The Impact of NASA on the Sunshine State* (Gainesville: University Press of Florida, 2002); David R. Goldfield, *Cotton Fields and Skyscrapers: Southern City and Region* (Baltimore: Johns Hopkins University Press, 1989).

114. Schulman, *From Cotton Belt to Sunbelt*, 135.

115. Daniel, "Going among Strangers," 898; Mark Schultz, *The Rural Face of White Supremacy* (Champaign: University of Illinois Press, 2005), 214.

116. Rew, "Annual Report," 1946, 4; M. S. Hoffman to J. B. Huston, January 31, 1946, Poultry, January 1–June 17, 1946, General Correspondence of the Office of the Secretary, 1906–70, Records of the Office of the Secretary of Agriculture, RG 16, NARA II.

117. J. B. Hutson to Albert W. Hawkes, January 23, 1946, Poultry, January 1–June 17, 1946, General Correspondence of the Office of the Secretary, 1906–70, Records of the Office of the Secretary of Agriculture, RG 16, NARA II.

118. Ibid.

119. J. B. Hutson to James M. Mead, February 7, 1946, 2, Poultry, January 1–June 17, 1946, General Correspondence of the Office of the Secretary, 1906–70, Records of the Office of the Secretary of Agriculture, RG 16, NARA II; Hutson to Hawkes, January 23, 1946.

120. James F. McGarvey to the U.S. Department of Agriculture, February 23, 1946, General Correspondence of the Office of the Secretary, 1906–70, Records of the Office of the Secretary of Agriculture, RG 16, NARA II.

121. Hutson to Mead, February 7, 1946; J. B. Hutson to Brian McMahon, March 5, 1946, Poultry, January 1–June 17, 1946, General Correspondence of the Office of the Secretary, 1906–70, Records of the Office of the Secretary of Agriculture, RG 16, 1.

122. Hutson to Mead, February 7, 1946.

123. Sanford Byers, interview by Lu Ann Jones, 29–30.

124. Ruby Byers, interview by Lu Ann Jones, April 23, 1987, 13, SAOHP.

125. H. A. Maxey, "Annual Narrative Report of Activities and Accomplishments in Cherokee County," 1946, 12, Annual Narrative and Statistical Reports, 1946–60, Records of the Extension Service, RG 33, NARA II.

126. Georgia Agricultural Extension Service, "Farming for Freedom: 1947 Annual Report," Georgia Agricultural Extension Service, University System of Georgia, 1948, 14; Maxey, "Annual Report," 1946, 11.

127. B. T. Brown, "Georgia Director's Report," 1948, 2, Annual Narrative and Statistical Reports, 1946–60, Records of the Extension Service, RG 33, NARA II; Kenneth Treanor, "Annual Report of Economics Extension Work, Project No. 7 for Calendar Year 1949," 1949, 25, Annual Narrative and Statistical Reports, 1946–60, Records of the Extension Service, RG 33, NARA II.

128. Maxey, "Annual Report," 1946, 11.

129. Arthur Gannon, "Annual Report Project No. 8-c Poultry," 1947, 4, Georgia Poultry 1947, Annual Narrative and Statistical Reports, 1946–60, RG 33, NARA II.

130. Helen Autry, "Annual Narrative of Activities and Accomplishments in Forsyth County," 1948, 3, Annual Narrative and Statistical Reports, 1946–60, Records of the Extension Service, RG 33, NARA II, 3; L. C. Rew, "Annual Narrative Report of Activities and Accomplishments in Hall County," 1948, 5, Hall County Georgia County Agent Annual Report 1948, Annual Narrative and Statistical Reports, 1946–60, RG 33, NARA II; John L. Anderson, "Annual Narrative Report of Activities and Accomplishments in Jackson County," 1946, 8, Jackson County, Georgia, County Agent Annual Report 1946, Annual Narrative and Statistical Reports, 1946–60, RG 33, NARA II; Maxey, "Annual Report," 1946, 11; Maxey, "Annual Report," 1947, 19.

131. Maxey, "Annual Report," 1947, 18; H. A. Maxey, "Annual Narrative Report of Activities and Accomplishments in Cherokee County," 1948, 20, Annual Narrative and Statistical Reports, 1946–60, Records of the Extension Service, RG 33, NARA II; H. A. Maxey, "Annual Narrative Report of Activities and Accomplishments in Cherokee County," 1949, 1, Annual

Narrative and Statistical Reports, 1946–60, Records of the Extension Service, RG 33, NARA II; Jeanette Harrell, "Annual Narrative Report of Activities and Accomplishments in Cherokee County," 1950, 22, Annual Narrative and Statistical Reports, 1946–60, Records of the Extension Service, RG 33, NARA II; Maxey, "Annual Report," 1950, 22.

132. Jack Temple Kirby, *Rural Worlds Lost: The American South, 1920–1960* (Baton Rouge: Louisiana State University Press, 1987).

Chapter 3. Taking Over

1. W. W. Harper and O. C. Hester, "Influence of Production Practices on Marketing of Georgia Broilers," Georgia Agricultural Experiment Stations, University of Georgia College of Agriculture (1956), 5; Packers and Stockyards Administration, "The Broiler Industry: An Economic Study of Structure, Practices and Problems" (Washington, D.C.: U.S. Department of Agriculture, 1967), 13.

2. "Cackle King from Georgia Sees L.A.," *Los Angeles Times*, March 13, 1952, 6, ProQuest Historical Newspapers.

3. Testimony of Jesse D. Jewell, May 13, 1957, Hearings before the Subcommittee No. 6 of the Select Committee on Small Business, House of Representatives, Eighty-Fifth Congress, First Session, 231, 217.

4. Arthur Gannon, "Annual Report Poultry Project No. 8-C for Calendar Year 1950," 1950, 5, Annual Narrative and Statistical Reports, 1946–60, Records of the Extension Service, RG 33, NARA II; John L. Anderson, "Annual Narrative Report of Activities and Accomplishments in Jackson County," 1950, 9, Annual Narrative and Statistical Reports, 1946–60, Records of the Extension Service, RG 33, NARA II.

5. Gannon, "Annual Report," 1950, 4.

6. Gannon, "Annual Report Poultry Project No. 8-C for Calendar Year 1949," 1949, 6–7, Annual Narrative and Statistical Reports, 1946–60, Records of the Extension Service, RG 33, NARA II.

7. Gannon, "Annual Report Poultry Project No. 8-C for Calendar Year 1951," 1951, 6, Annual Narrative and Statistical Reports, 1946–60, Records of the Extension Service, RG 33, NARA II.

8. U.S. Department of Agriculture, "Financing Production and Marketing of Broilers in the South: Part I, Dealer Phase," Southern Cooperative Series Bulletin no. 38 (June 1954), Agricultural Experiment Stations of Alabama, Arkansas, Georgia, Louisiana, Mississippi, North Carolina, South Carolina, Tennessee, Texas, and Virginia, and the Agricultural Marketing Service, 47; W. W. Harper and O. C. Hester, "Influence of Production Practices on Marketing of Georgia Broilers," Georgia Agricultural Experiment Stations, University of Georgia College of Agriculture (1956), 9; W. W. Harper, "Marketing Georgia Broilers," Bulletin no. 281 (July 1953), University of Georgia College of Agriculture Experiment Stations, 23, 13–14, 13, 28.

9. Harper and Hester, "Influence of Production Practices," 9; Harper, "Marketing Georgia Broilers," 13, 28.

10. Brent Riffel, "The 'Nuevo South': Tyson Foods and the Transformation of American Labor," *Southern Historian* 29 (Spring 2008): 22.

11. John O. Gerald and Humbert S. Kahle, "Marketing Georgia Broilers through Commercial Processing Plants," Agricultural Marketing Service and Georgia Experiment Station,

1955, 8, 1, i; "Hall Poultry Area May Feed the 'Meatless,'" undated newspaper clipping pasted in L. C. Rew, "Annual Narrative Report of Activities and Accomplishments in Hall County," 1947, 5–6, unpaginated pages of clippings, Hall County Georgia County Agent Annual Report, 1947, Annual Narrative and Statistical Reports, 1946–60, Records of the Extension Service, RG 33, NARA II.

12. Gerald and Kahle, "Marketing Georgia Broilers."

13. John L. Anderson, "Annual Narrative Report of Activities and Accomplishments in Jackson County," 1946, 2, Annual Narrative and Statistical Reports, 1946–60, Records of the Extension Service, RG 33, NARA II; John L. Anderson, "Annual Narrative Report of Activities and Accomplishments in Jackson County," 1947, 2, Annual Narrative and Statistical Reports, 1946–60, Records of the Extension Service, RG 33; "Hall Poultry Area May Feed the 'Meatless,'" undated newspaper clipping in Rew, "Annual Narrative Report," 1947, unpaginated pages of clippings.

14. L. C. Rew, "Annual Narrative Report of Activities and Accomplishments in Hall County," 1946, 5, Annual Narrative and Statistical Reports, 1946–60, Records of the Extension Service, RG 33, NARA II.

15. "Poultry Operators Study New Package Methods of Experts," undated newspaper clipping in Rew, "Annual Narrative Report," 1947, unpaginated clippings; Rew, "Annual Narrative Report," 1947, 23; and L. C. Rew, "Annual Narrative Report of Activities and Accomplishments in Hall County," 1948, 16, Annual Narrative and Statistical Reports, 1946–60, Records of the Extension Service, RG 33, NARA II.

16. Testimony of Jewell, May 13, 1957, 231.

17. "Poultry Operators Study New Package Methods," 5.

18. O. C. Hester and W. W. Harper, "The Function of Feed-Dealer Suppliers in Marketing Georgia Broilers," Bulletin no. 283 (August 1953), University of Georgia College of Agriculture Experiment Stations in cooperation with the Bureau of Agricultural Economics, U.S. Department of Agriculture, 30.

19. Harper, "Marketing Georgia Broilers," 10.

20. Peter L. Hansen and Ronald L. Mighell, "Economic Choices in the Broiler Industry" (Washington, D.C.: U.S. Department of Agriculture, 1956), 20–21.

21. Robert B. Powers, Richard K. Noles, and James C. Fortson, "Broiler Production: A Study of the Growing Operation" (Athens: University of Georgia, College of Agriculture Experiment Stations, 1967), 5.

22. Gilbert C. Fite, *Cotton Fields No More: Southern Agriculture, 1865–1980* (Lexington: University Press of Kentucky, 1984), 201; Gordon Sawyer, *The Agribusiness Poultry Industry: A History of Its Development* (New York: Exposition Press, 1971), chap. 6.

23. Harper, "Marketing Georgia Broilers," 26, 10.

24. Testimony of Jewell, May 13, 1957, 217.

25. Ibid., 233.

26. U.S. Congress, House Select Committee on Small Business, "Small Business Problems in the Poultry Industry: A Report of the Select Committee on Small Business, House of Representative, Eighty-Seventh Congress, Second Session, Pursuant to H. Res. 46" (Washington, D.C.: U.S. Government Printing Office, 1963), 4–5.

27. Harper, "Marketing Georgia Broilers," 11.

28. Hester and Harper, "Function of Feed-Dealer Suppliers," 31.

29. Harper, "Marketing Georgia Broilers," 14.

30. Harper and Hester, "Influence of Production Practices," 9–10.

31. Ibid.; Packers and Stockyards Administration, "The Broiler Industry: An Economic Study of Structure, Practices and Problems" (Washington, D.C.: U.S. Department of Agriculture, 1967), 10.

32. Harper and Hester, "Influence of Production Practices," 10.

33. Testimony of Jewell, May 13, 1957, 233.

34. Arthur Flemming, interview by Lu Ann Jones, April 27, 1987, 65–66, Southern Agriculture Oral History Project, National Museum of American History, Washington, D.C.

35. U.S. Congress, House Select Committee on Small Business, "Small Business Problems," 22.

36. Harper and Hester, "Influence of Production Practices," 9; Harper, "Marketing Georgia Broilers," 12.

37. Hansen and Mighell, "Economic Choices."

38. Douglas R. Constance, "The Southern Model of Broiler Production and Its Global Implications," *Culture & Agriculture* 30, nos. 1–2 (2008): 17–31.

39. Testimony of Jewell, May 13, 1957, 228, 229.

40. U.S. Congress, House Select Committee on Small Business, "Small Business Problems," 10.

41. W. S. Brown, "1951 Annual Report," Georgia Director's Report 1951, 11, Annual Narrative and Statistical Reports, 1946–60, Records of the Extension Service, RG 33, NARA II; S. G. Chandler, "Annual Report North Georgia District for Calendar Year 1951," 1951, 73, Annual Narrative and Statistical Reports, 1946–60, Records of the Extension Service, RG 33, NARA II; Testimony of Jewell, May 13, 1957, 220; Testimony of Oscar Straube, President, Pay Way Feed Mills, Inc., and Chairman of the Board, American Feed Manufacturers Association, Problems in the Poultry Industry, Hearings before the Subcommittee No. 6 of the Select Committee on Small Business, House of Representatives, Eighty-Fifth Congress, First Session, May 14, 1957, 4; H. A. Maxey, "Annual Narrative Report of Activities and Accomplishments in Cherokee County," 1955, 4, Annual Narrative and Statistical Reports, 1946–60, Records of the Extension Service, RG 33, NARA II; Bernard F. Tobin and Henry Arthur, *Dynamics of Adjustment in the Broiler Industry* (Boston: Harvard University Press, 1964); Ross Talbot, *The Chicken War: An International Trade Conflict between the United States and the European Economic Community, 1961–1964* (Ames: Iowa State University Press, 1978); U.S. Congress, House Select Committee on Small Business, "Small Business Problems"; Arthur Curran, "The Georgia Broiler Industry: A Case Study of Evolving Industrial Patterns in an Industry over a Decade" (PhD dissertation, University of Georgia, 1970), 53.

42. Testimony of Jewell, May 13, 1957, 230, 236.

43. L. C. Rew, "Annual Narrative Report of Activities and Accomplishments in Hall County," 1949, 23, Annual Narrative and Statistical Reports, 1946–60, Records of the Extension Service, RG 33, NARA II; L. C. Rew, "Annual Narrative Report of Activities and Accomplishments in Hall County," 1950, 23, Annual Narrative and Statistical Reports, 1946–60, Records of the Extension Service, RG 33, NARA II.

44. Zelma R. Bannister, "Annual Narrative Report of Activities and Accomplishments in Cherokee County, Forsyth, Georgia, Home Demonstration Agent Annual Report," 1951, 3, Annual Narrative and Statistical Reports, 1946–60, Records of the Extension Service, RG 33, NARA II.

45. H. A. Maxey, "Annual Narrative Report of Activities and Accomplishments in Cherokee County," 1951, 19, Annual Narrative and Statistical Reports, 1946–60, Records of the Extension Service, RG 33, NARA II.

46. B. T. Brown, "Forsyth County Narrative Report," 1951, 2, Extension Service Annual Reports, Georgia, Annual Narrative and Statistical Reports, 1908–1974, Records of the Extension Service, RG 33, NARA II.

47. John L. Anderson, "Narrative Annual Report of Extension Work Conducted in Jackson County, Georgia," 1951, 21, Extension Service Annual Reports, Annual Narrative and Statistical Reports, 1908–1974, Records of the Extension Service, RG 33, NARA II.

48. Packers and Stockyards Administration, "Broiler Industry," 13.

49. Testimony of Jewell, May 13, 1957, 2.

Chapter 4. Broiler Sharecroppers and Hired Hands

1. Arthur Flemming, interview by Lu Ann Jones, April 21, 1987, 17, 21, 37, 38, 46, Southern Agriculture Oral History Project, National Museum of American History, Washington, D.C. (hereafter SAOHP).

2. R. L. Kohls and J. W. Wiley, "Aspects of Multiple-Owner Integration in the Broiler Industry," *Journal of Farm Economics* 37, no. 1 (1955): 81.

3. W. W. Harper and O. C. Hester, "Influence of Production Practices on Marketing of Georgia Broilers," Georgia Agricultural Experiment Stations, University of Georgia College of Agriculture (1956), 5, 25; Wallace Hugh Warren, "Progress and Its Discontents: The Transformation of the Georgia Foothills, 1920–1970" (master's thesis, University of Georgia, 1997), 122–23.

4. Harper and Hester, "Influence of Production Practices," 25.

5. Reese Cleghorn, "Appalachia—Poverty, Beauty, and Poverty," *New York Times*, April 25, 1965, accessed January 12, 2012, ProQuest Historical Newspapers.

6. Sanford Byers, interview by Lu Ann Jones, April 20, 1987, 33, SAOHP.

7. Packers and Stockyards Administration, "The Broiler Industry: An Economic Study of Structure, Practices and Problems" (Washington, D.C.: U.S. Department of Agriculture, 1967), 13.

8. W. W. Harper, "Marketing Georgia Broilers," Bulletin no. 281 (July 1953), University of Georgia College of Agriculture Experiment Stations, 1953, 31.

9. Spurgeon Welborn, interview by Lu Ann Jones, April 27, 1987, 40–42, SAOHP.

10. Flemming, interview by Lu Ann Jones, April 21, 1987, 31, 33, SAOHP.

11. Testimony of Jesse D. Jewell, May 13, 1957, Hearings before the Subcommittee No. 6 of the Select Committee on Small Business, House of Representatives, Eighty-Fifth Congress, First Session, 227.

12. Arthur Flemming, interview by Lu Ann Jones, April 27, 1987, 66, SAOHP.

13. Ruby Byers, interview by Lu Ann Jones, April 27, 1987, 37, SAOHP.

14. Packers and Stockyards Administration, "Broiler Industry," 14.

15. Everett O. Stoddard, "Costs and Returns: Broiler Farms, Georgia, 1962," in *The Poultry and Egg Situation* (Washington, D.C.: Economic Research Service, U.S. Department of Agriculture, 1964), 14–20; Walter H. Rucker, "Annual Narrative Report of Activities and Accomplishments in Forsyth County," 1956, Annual Narrative and Statistical Reports, 1946–60, Records of the Extension Service, RG 33, NARA II; John L. Anderson, "Annual Narrative Report of Activities and Accomplishments in Jackson County," 1951, 10, Annual Narrative and Statistical Reports, 1946–60, Records of the Extension Service, RG 33, NARA II; W. S. Brown, "Georgia Director's Report," 1950, 12, Annual Narrative and Statistical Reports, 1946–60, Records of the Extension Service, RG 33, NARA II.

16. Richard K. Noles and Milton Y. Dendy, "Broiler Production in Georgia: Grower's Costs and Returns," University of Georgia College of Agriculture Experiment Stations, 1968, 5.

17. Harper, "Marketing Georgia Broilers," 31–32.

18. Ibid., 32.

19. Sanford Byers, interview by Lu Ann Jones, 33, 34.

20. O. C. Hester and W. W. Harper, "The Function of Feed-Dealer Suppliers in Marketing Georgia Broilers," Bulletin no. 283 (August 1953), University of Georgia College of Agriculture Experiment Stations in cooperation with the Bureau of Agricultural Economics, U.S. Department of Agriculture, 9–11.

21. Flemming, interview by Lu Ann Jones, April 27, 1987, 65–66, SAOHP.

22. Welborn, interview by Lu Ann Jones, 42–43.

23. Ibid., 43.

24. Flemming, interview by Lu Ann Jones, April, 27, 1987, 65.

25. Hester and Harper, "Function of Feed-Dealer Suppliers," 10.

26. Flemming, interview by Lu Ann Jones, April 27, 1987, 67.

27. Welborn, interview by Lu Ann Jones, 42; D. W. Brooks, "Meeting the Challenge of Vertical Integration," speech before the American Institute of Cooperation, Ft. Collins, Colo., August 20, 1957, David Walker Brooks Collection, Richard B. Russell Library for Political Research and Studies, University of Georgia Libraries, Athens, Ga.; David W. Brooks, "Agricultural Changes," speech given before the Federal Land Bank, Houston, Tex., January 22, 1959, David Walker Brooks Collection, Richard B. Russell Library for Political Research and Studies, University of Georgia Libraries, Athens, Ga.; U.S. Department of Agriculture, "Financing Production and Marketing of Broilers in the South: Part I, Dealer Phase," Southern Cooperative Series Bulletin no. 38 (June 1954), Agricultural Experiment Stations of Alabama, Arkansas, Georgia, Louisiana, Mississippi, North Carolina, South Carolina, Tennessee, Texas, and Virginia, and the Agricultural Marketing Service, 62; Ray Marshall and Allen Thompson, "Status and Prospects of Small Farmers in the South," Southern Regional Council Report, Atlanta, Ga., 1976, 55; William Boyd and Michael Watts, "Agro-Industrial Just-in-Time: The Chicken Industry and Postwar American Capitalism," in *Globalising Food: Agrarian Questions and Global Restructuring*, ed. David Goodman and Michael Watts (London: Routledge, 1997), 199–213, 211.

28. Marshall and Thompson, "Status and Prospects of Small Farmers," 55.

29. Welborn, interview by Lu Ann Jones, 24.

30. Georgia Agricultural Extension Service, "Farming for Freedom: 1947 Annual Report," Georgia Agricultural Extension Service, University System of Georgia, 1948, 3.

31. John L. Anderson, "Annual Narrative Report of Activities and Accomplishments in Jackson County," 1947, 20, Annual Narrative and Statistical Reports, 1946–60, Records of the Extension Service, RG 33, NARA II.

32. "Rew Urges Buying of Broilers to Bolster Market as Prices Fall: County Agent Says Process below Production Cost; Freezer Owners Can Help Prevent Drop," undated newspaper clipping pasted into L. C. Rew, "Annual Narrative Report of Activities and Accomplishments in Hall County," 1947, Annual Narrative and Statistical Reports, 1946–60, Records of the Extension Service, RG 33, NARA II.

33. John L. Anderson, "Annual Narrative Report of Activities and Accomplishments in Jackson County," 1948, 19, Annual Narrative and Statistical Reports, 1946–60, Records of the Extension Service, RG 33, NARA II; B. T. Brown, "Annual Report," 1948, 1–2, Annual

Narrative and Statistical Reports, 1946–60, Records of the Extension Service, RG 33, NARA
II; Kenneth Treanor, "Annual Report of Economics Extension Work, Project No. 7 for
Calendar Year 1949," 1949, 25, Annual Narrative and Statistical Reports, 1946–60, Records
of the Extension Service, RG 33, NARA II.

34. Treanor, "Annual Report," 1949, 4.

35. H. A. Maxey, "Annual Narrative Report of Activities and Accomplishments in Cherokee County," 1946, 12, Annual Narrative and Statistical Reports, 1946–60, Records of the
Extension Service, RG 33, NARA II.

36. Stoddard, "Costs and Returns," 14.

37. Testimony of Jewell, May 13, 1957, 220, 224.

38. H. A. Maxey, "Annual Narrative Report of Activities and Accomplishments in Cherokee County," 1955, 27, Annual Narrative and Statistical Reports, 1946–60, Records of the
Extension Service, RG 33, NARA II; U.S. Congress, House Select Committee on Small Business, "Small Business Problems in the Poultry Industry: A Report of the Select Committee
on Small Business, House of Representatives, Eighty-Seventh Congress, Second Session,
Pursuant to H. Res. 46" (Washington, D.C.: U.S. Government Printing Office, 1963), 5.

39. Marshall and Thompson, "Status and Prospects of Small Farmers," 55.

40. Testimony of Jewell, May 13, 1957, 224.

41. Gilbert C. Fite, *Cotton Fields No More: Southern Agriculture, 1865–1980* (Lexington:
University Press of Kentucky, 1984), 201; Gordon Sawyer, *The Agribusiness Poultry Industry:
A History of Its Development* (New York: Exposition Press, 1971), chap. 6; Harper, "Mar-
keting Georgia Broilers," 10–25; Hester and Harper, "Function of Feed-Dealer Suppliers,"
18–25; Testimony of Jewell, May 13, 1957, 217; Hester and Harper, "Function of Feed-Dealer
Suppliers," 30.

42. B. T. Brown, "Forsyth County Narrative Report," 1951, 2, Extension Service Annual
Reports, Georgia, Annual Narrative and Statistical Reports, 1908–1974, Records of the
Extension Service, RG 33, NARA II; John L. Anderson, "Jackson Georgia County Agent
Annual Report," 1951, 21, Annual Narrative and Statistical Reports, 1908–1974, Records of
the Extension Service, RG 33, NARA II; H. A. Maxey, "Annual Narrative Report of Activi-
ties and Accomplishments in Cherokee County," 1954, 19, Annual Narrative and Statistical
Reports, 1946–60, Records of the Extension Service, RG 33, NARA II; H. A. Maxey, "Annual
Narrative Report of Activities and Accomplishments in Cherokee County," 1955, 2, Annual
Narrative and Statistical Reports, 1946–60, Records of the Extension Service, RG 33, NARA
II; H. A. Maxey, "Annual Narrative Report of Activities and Accomplishments in Cherokee
County," 1956, 2, Annual Narrative and Statistical Reports, 1946–60, Records of the Exten-
sion Service, RG 33, NARA II.

43. Rucker, "Annual Narrative Report," 1956.

44. Maxey, "Annual Narrative Report," 1954, 1; L. R. Lanier, "Annual Narrative Report
of Extension Supervisors (Men) State of Georgia," 1954, 16, Annual Narrative and Statistical
Reports, 1946–60, Records of the Extension Service, RG 33, NARA II.,

45. Testimony of Jewell, May 13, 1957, 217; Flemming, interview by Lu Ann Jones,
April 21, 1987, 28, SAOHP; Noles and Dendy, "Broiler Production in Georgia"; Clay Fulcher,
"Vertical Integration in the Poultry Industry: The Contractual Relationship," *Agricultural
Law Update* (January 1992): 4–6; E. Roy, *Contract Farming and Economic Integration* (Dan-
ville, Ill.: Interstate Printers & Publishers, 1972); J. M. Sprott and H. Jackson, "Contract
Broiler Growers in Arkansas," *Arkansas Farm Research* (1964): 11; Welborn, interview by Lu
Ann Jones, 40; Ruby Byers, interview by Lu Ann Jones, Apr. 23, 1987, transcript, 7, SAOHP.

46. U.S. Department of Agriculture, "Financing Production and Marketing of Broilers in the South: Part I, Dealer Phase," Southern Cooperative Series Bulletin no. 38 (June 1954), Agricultural Experiment Stations of Alabama, Arkansas, Georgia, Louisiana, Mississippi, North Carolina, South Carolina, Tennessee, Texas, and Virginia, and the Agricultural Marketing Service, 62.

47. Welborn, interview by Lu Ann Jones, 43.

48. Walter H. Rucker, "Annual Narrative Report of Activities and Accomplishments in Forsyth County," 1954, 10, Annual Narrative and Statistical Reports, 1946–60, Records of the Extension Service, RG 33, NARA II.

49. Rucker, "Annual Narrative Report," 1956.

50. Harper and Hester, "Influence of Production Practices," 25; Stoddard, "Costs and Returns," 14–20.

51. L. C. Rew, "Annual Narrative Report of Activities and Accomplishments in Hall County," 1955, 1, Annual Narrative and Statistical Reports, 1946–60, Records of the Extension Service, RG 33, NARA II; H. A. Maxey, "Annual Narrative Report of Activities and Accomplishments in Cherokee County," 1956, 21, Annual Narrative and Statistical Reports, 1946–60, Records of the Extension Service, RG 33, NARA II; Louise Kemp, "Annual Narrative Report of Activities and Accomplishments in Cherokee County," 1953, 1, Annual Narrative and Statistical Reports, 1946–60, Records of the Extension Service, RG 33, NARA II; Walter Rucker, "Annual Narrative Report of Activities and Accomplishments in Forsyth County," 1953, 2, Annual Narrative and Statistical Reports, 1946–60, Records of the Extension Service, RG 33, NARA II; Flemming, interview by Lu Ann Jones, April 21, 1987, 47; Flemming, interview by Lu Ann Jones, April 27, 1987, 59; Frances A. Eubanks, "Annual Narrative Report of Activities and Accomplishments in Cherokee County," 1954, Annual Narrative and Statistical Reports, 1946–60, Records of the Extension Service, RG 33, NARA II.

52. Testimony of Jewell, May 13, 1957, 217, 223–24, 228.

53. "Cackle King from Georgia Sees L.A.," Los Angeles Times, March 13, 1952, 6, ProQuest Historical Newspapers.

54. "Ask the Times: Poultry Statue Went Up in 1977," Gainesville Times, August 24, 2012, accessed March 2, 2013, http://www.gainesvilletimes.com/archives/71835/.

55. Testimony of Jewell, May 13, 1957, 238; Paul Simmons, "Chicken Industry Revolutionizing Lives of Thousands of Georgians," Rome News-Tribune, April 28, 1957; H. A. Maxey, "Annual Narrative Report of Activities and Accomplishments in Cherokee County," 1950, 22, Annual Narrative and Statistical Reports, 1946–60, Records of the Extension Service, RG 33, NARA II.

56. Flemming, interview by Lu Ann Jones, April 21, 1987, 47; Flemming, interview by Lu Ann Jones, April 27, 1987, 59.

57. H. A. Maxey, "Annual Narrative Report of Activities and Accomplishments in Cherokee County," 1951, 1, 19, 23, Annual Narrative and Statistical Reports, 1946–60, Records of the Extension Service, RG 33, NARA II; Sara A. Vanhorn, "Annual Narrative Report of Activities and Accomplishments in Hall County," 7, Hall Georgia Home Demonstration Agent Annual Report 1952, Annual Narrative and Statistical Reports, 1946–60, Records of the Extension Service, RG 33, NARA II.

58. Packers and Stockyards Administration, "Broiler Industry," 20, iv; Stoddard, "Costs and Returns," 14–21.

59. Flemming, interview by Lu Ann Jones, April 21, 1987, 21, 17, 37, 38.

Chapter 5. From Public Nuisance to Toxic Waste, 1940–1990

1. James Aswell, "Chickens in the Wind," *Colliers,* September 9, 1950, 31.

2. W. W. Harper and O. C. Hester, "Influence of Production Practices on Marketing of Georgia Broilers," Georgia Agricultural Experiment Stations, University of Georgia College of Agriculture (1956), 5.

3. Willard Range, *A Century of Georgia Agriculture, 1850–1950* (Athens: University of Georgia Press, 1954), 201.

4. Hollis Shomo, Gordon Tucker, Cecil Rogers, and Paul Abbey, *The Broiler Industry: In Georgia and Other States in the Southeast* (Richmond: Virginia Department of Agriculture, Division of Markets, 1955), ii.

5. Velva Blackstock, interview by Lu Ann Jones, April 22, 1987, 60, Southern Agriculture Oral History Project, National Museum of American History, Washington, D.C. (hereafter SAOHP).

6. Ibid., 60.

7. Harper and Hester, "Influence of Production Practices," 29–30.

8. *W. C. Strickland et al. v. Douglas Lambert* (1959), No. 8 Div. 938, 268 Ala. 580; 109 So. 2d 664; 1959 Ala. LEXIS 371.

9. H. A. Maxey, "Annual Narrative Report of Activities and Accomplishments in Cherokee County," 1952, 4, Annual Narrative and Statistical Reports, 1946–60, Records of the Extension Service, RG 33, NARA II.

10. County Agent's Column, *Upcountry (Canton) Georgia Tribune,* undated article pasted into Maxey, "Annual Report," 1952, 4; W. W. Harper, "Marketing Georgia Broilers," Bulletin no. 281 (July 1953), University of Georgia College of Agriculture Experiment Stations, 1953, 21.

11. William Cronon, *Nature's Metropolis: Chicago and the Great West* (New York: Norton, 1992), 208–9.

12. L. C. Rew and W. V. Chafin, "Annual Report Hall County, Georgia," 1943, 5, microfilm roll 133, Extension Service Annual Reports, Georgia, 1909–44, Annual Narrative and Statistical Reports, 1908–1974, Records of the Extension Service, RG 33, NARA II; John O. Gerald and Humbert S. Kahle, "Marketing Georgia Broilers through Commercial Processing Plants," Agricultural Marketing Service and Georgia Experiment Station, 1955, 23 and 24–26]]; *Clifford E. Wood v. Dorsey W. Sutton et al.* (1975), 2 Va. Cir. 425; 1975 Va. Cir. LEXIS 4.

13. Testimony of Jesse D. Jewell, May 13, 1957, Hearings before the Subcommittee No. 6 of the Select Committee on Small Business, House of Representatives, Eighty-Fifth Congress, First Session, 217.

14. J. B. Ruhl, "Farms, Their Environmental Harms, and Environmental Law," *Ecology Law Quarterly* 27, no. 263 (2000): 315.

15. *Poultryland Inc. v. Anderson et al.* (1946), No. 15398, Supreme Court of Georgia, 200 Ga. 549; 37 S.E.2d 785; 1946 Ga. LEXIS 302, April 2, 1946, Decided.

16. *Green v. Smith* (1959), No. 5-1987, 231 Ark. 94; 328 S.W.2d 357; 1959 Ark. LEXIS 468.

17. *Colley et al. v. Tatum et al.* (1971), No. 26227, 227 Ga. 294, 180 S.E.2d 346, 1971 Ga. LEXIS 675, 2 ERC (BNA) 1327; *O. L. Traylor, Sr., et al., Plaintiff-Appellant v. Artis Colvin, Defendant-Appellee* (1955), No. 8411, 84 So. 2d 286, 1955 La. App. LEXIS 1068; *Ozark Poultry Products, Inc. v. Roy Garman et al.* (1971), No. 5-5647, 251 Ark. 389; 472 S.W.2d 714; 1971 Ark. LEXIS 1151; 3 ERC (BNA) 1545; 2 ELR 20016.

18. *Ozark Bi-Products, Inc. v. Bohannon* (1954), No. 5-435, 224 Ark. 17, 271 S.W.2d 354, 1954 Ark. LEXIS 514.

19. *Bishop Processing Company v. Davis et al.* (1957), No. 223, October Term, 1956, 213 Md. 465, 132 A.2d 445, 1957 Md. LEXIS 607.

20. *Alfred Jacobshagen Company Trading as Jackson Reduction Company v. Dockery, et al.* (1962), No. 42197, 243 Miss. 511, 139 So. 2d 632, 1962 Miss. LEXIS 369.

21. *Irby et al. v. Lamb et al.* (1963), No. 21994, 218 Ga. 840; 131 S.E.2d 183; 1963 Ga. LEXIS 344.

22. *Green v. Smith* (1959).

23. Kelly A. Reynolds, Kristina D. Mena, and Charles P. Gerba, "Risk of Waterborne Illness via Drinking Water in the United States," *Reviews of Environmental Contamination and Toxicology* 192 (2008): 117–58.

24. Maryn McKenna, "News: FDA Won't Act Against Ag Antibiotic Use," *Wired*, December 23, 2011, accessed July 16, 2015, http://www.wired.com/2011/12/fda-ag-antibiotics/; Maryn McKenna, "Report: FDA Documents Show Decade of Unsuccessful Attempts to Control Farm Antibiotics," Wired, January 28, 2014, accessed July 16, 2015, http://www .wired.com/2014/01/nrdc-report-fda/; Natural Resources Defense Council (NRDC), "Playing Chicken with Antibiotics: Previously Undisclosed FDA Documents Show Antibiotic Feed Additives Don't Meet the Agency's Own Safety Standards," January 27, 2014, accessed July 16, 2015, http://www.nrdc.org/food/saving-antibiotics/antibiotic-feed-fda-documents .asp; NRDC, "It's Time for the U.S. Poultry Industry to Demonstrate Antibiotic Stewardship," May 23, 2014, accessed July 16, 2015, http://www.nrdc.org/food/saving-antibiotics /poultry-industry-antibiotic-stewardship.asp.

25. Charles Duhigg, "Health Ills Abound as Farm Runoff Fouls Wells," *New York Times*, September 17, 2009, accessed July 16, 2015, http://www.nytimes.com/2009/09/18/us/18dairy .html?pagewanted=all.

26. See Josh Marks, "Regulating Agricultural Pollution in Georgia: Recent Trends and the Debate over Integrator Liability," *Georgia State University Law Review* 18 (Summer 2002): 1031; Rebecca P. Lewandoski, "Spreading the Liability Net: Overcoming Agricultural Exemption with EPA's Proposed Co-permitting Regulation under the Clean Water Act," *Vermont Law Review* 27 (Fall 2002): 149; Cynthia M. Roelle, "Pork, Pollution, and Priorities: Integrator Liability in North Carolina," *Wake Forest Law Review* 35 (Winter 2000): 1055.

27. On agribusiness as a leading source of water pollution, see Marks, "Regulating Agricultural Pollution in Georgia," 1036; Dana R. Flick, "Comment: The Future of Agricultural Pollution Following USDA and EPA Drafting of a Unified National Strategy for Animal Feeding Operations," *Dickinson Journal of Environmental Law and Policy* 8 (1999): 61. EPA studies find that farming is the leading pollution source in 60 percent of rivers and 30 percent of lakes. U.S. Environmental Protection Agency, *The Quality of Our Nation's Waters: A Summary of the National Water Quality Inventory*, 1998 Report to Congress (2000), 6–9; U.S. Environmental Protection Agency, *Strategy for Addressing Environmental and Public Health Impacts from Animal Feeding Operations*, proposed March 4, 1998, 1–2.

28. Josh Marks, "Regulating Agricultural Pollution in Georgia," 1031; Lewandoski, "Spreading the Liability Net," 149; Roelle, "Pork, Pollution, and Priorities," 1055.

29. Harper, "Marketing Georgia Broilers," 34–35; Richard K. Noles and Milton Y. Dendy, "Broiler Production in Georgia: Grower's Costs and Returns," University of Georgia College of Agriculture Experiment Stations, 1968, 15, 6; Stephen J. Brannen, "An Analysis

of the Financing of Broiler Production in North Georgia" (master's thesis, University of Georgia, 1952), 38; Harper and Hester, "Influence of Production Practices," 21, 29; U.S. Department of Agriculture, "Financing Production and Marketing of Broilers in the South: Part I, Dealer Phase," Southern Cooperative Series Bulletin no. 38 (June 1954), Agricultural Experiment Stations of Alabama, Arkansas, Georgia, Louisiana, Mississippi, North Carolina, South Carolina, Tennessee, Texas, and Virginia, and the Agricultural Marketing Service, 14; O. C. Hester and W. W. Harper, "The Function of Feed-Dealer Suppliers in Marketing Georgia Broilers," Bulletin no. 283 (August 1953), University of Georgia College of Agriculture Experiment Stations in cooperation with the Bureau of Agricultural Economics, U.S. Department of Agriculture, 11.

30. Gregory W. Blount, Douglas A. Henderson, and Debra S. Cline, "The New Nonpoint Source Battleground: Concentrated Animal Feeding Operations" *Natural Resources & Environment* 14, no. 1 (1999): 42–45, 68–69.

31. Duhigg, "Health Ills Abound."

32. Chris Joyner, "Fecal Risk to Water Found," *Atlanta Journal-Constitution*, August 14, 2011, accessed July 16, 2016, http://www.ajc.com/news/news/local/fecal-risk-to-water-found /nQKjW/.

33. Ibid.

Epilogue

1. Robert Tuten and David Amey, "Reaching for the Rising Sun," *Broiler Industry*, June 1989, 29.

2. Doreen Carvajal and Stephen Castle, "A U.S. Hog Giant Transforms Eastern Europe," *New York Times*, May 6, 2009, accessed July 16, 2015, http://www.nytimes.com/2009/05/06 /business/global/06smithfield.html?_r=1.

3. Associated Press, "Tyson Looks Abroad," reprinted in *New York Times*, November 13, 2007, accessed July 16, 2015, http://www.nytimes.com/2007/11/13/business/13tyson.html?scp =1&sq=tyson%20looks%20abroad&st=cse.

4. Tuten and Amey, "Reaching for the Rising Sun," 28.

5. Ibid., 28.

6. William D. Heffernan and Douglas H. Constance, "Transnational Corporations and the Globalization of the Food System," in *From Columbus to Con Agra: The Globalization of Agriculture and Food*, ed. Alessandro Bonanno, Lawrence Busch, William Friedland, Lourdes Gouveia, and Enxo Mingione (Kansas: University of Kansas Press, 1994), 43.

7. Gary Vocke, "Investment to Transfer Poultry Production to Developing Countries," *American Journal of Agricultural Economics* 73, no. 3 (August 1991): 951–54.

8. Carvajal and Castle, "U.S. Hog Giant Transforms Eastern Europe."

INDEX

National Poultry Improvement Plan
(1935), 20
Negro Home Demonstration Agents, 32
New Deal, xi, 10–14, 33
Nippon Meat Packers of Japan, 71
no-loss provision contract, 43–45
Northeast-Georgia Fair, 8
nuisance suits, 65–68
nutritional deficiencies, 20

offal, 65, 67
one-crop economy: cotton, 5; poultry
industry replaces cotton, x, xi, xii, 2–3,
24, 36–37, 50, 51, 54, 61
open account, 43–44
Ozarks, xii

pellagra, 24
pesticides, 2
point source pollution, 68
poultry industry: as agriculture versus
industry, 19; confinement of chickens,
19–20; congressional hearings on, 47,
49, 57; control, farmers' lack of, xii,
38–39, 42–43, 47, 50; cotton farming,
comparison with, 59; cotton replaced
by, x, xi, xii, 2–3, 24, 36–37, 50, 51, 61;
crop lien system, 2–3, 6, 21, 22, 44;
diseases, research on, 20, 23, 31–32,
40–41; diversification, lack of, 21–22;
Georgia economy, 3; globalization,
71–72; government-funded research, ix,
20, 23, 31–32, 40–41; growth of, 15, 17,
19, 21, 23, 63–65, 67; history of poultry
farming, 15–16; income and, 32, 37,
39, 43, 51, 56–59 (see also wage work);
industrialization of agriculture, 19, 36;
investments, 16–17, 23–24, 36, 38, 43,
52, 57; location of, 1; markets, 18–19,
48; prices, 48; as seasonal product, 18–
19; small-scale farmers pushed out, 51–
52, 54; standardization, 39; surpluses,
35; waste, 63–69
Poultryland Inc. See Jewell, Jesse Dixon
processing plants, ix–x, 30–32, 41, 65
public work. See wage work

Puckette, Charles, 11
pullorum, 20
Purina, 14, 60

quasi-vertical integration, x–xii, 2, 39–42

railroad car sales, 18
railroads, 6
Raper, Arthur, 12, 13
rendering plants, 65, 67
Rew, L. C., 22, 32, 65
Roosevelt, Franklin D., 12
Rural Electrification Administration, 19

satellite activities, 46–48
Sawyer, Gordon, 77n49
Sears and Roebuck, 16
"shakedown," 23
sharecropping, 24; African American
farmers and, 32; Agricultural Adjustment Administration and, 13–14;
chicken farmers as tenants, 56; decline
of, 25–26, 33, 34; industrialization of
agriculture and, 26–27; rise of, 5, 7; soil
exhaustion and, 7; tenancy, decline of,
13; tenants and cotton allotments,
12–13
Sherman, William Tecumseh, 6
small farmers: cotton allotments, 13; forced
out, 27, 39, 51–52, 54; "shakedown"
and, 23
Smith, Ruby Faye, 16, 17
Smithfield Farms, 71, 72
Soil Conservation and Domestic Allotment Act (1936), 12
soil erosion, ix, 5, 24
soil exhaustion, 7, 24
Soule, Andrew M., 5
Steagall Amendment (1941), 28–29
Stephenson, J. S., 15
Street, Allie, 16
Sun Valley Thailand, 71

Tatum, B. A., 51–52, 54
tenancy. See sharecropping
Trasgo Group, 71

Treanor, Kenneth, 27
Tyson, John, 41
Tyson Foods, 41, 71

urbanization, 35
USDA: African American farmers, exclusion
    of, 3; cotton and, 8, 10–14; on farmers'
    upgrades, 46; feed conversion contract,
    definition of, 42; hatcheries and, 30; on
    industrialization of agriculture, 26–27;
    no-loss provision contract, 44–45; price
    setting, 29–30; small-scale farmers and,
    33–34; on wartime poultry production,
    28–30

Vance, Rupert, 10, 11, 22

wage work, 50, 51, 59–61
War Food Administration, 28–30
War Food Order 119, 29

War Food Order 142, 30
Weems, Camilla, 32
Welborn, Spurgeon: on bartering, 16; on
    farmers as tenants, 56; on industriali-
    zation of agriculture, 27; on lack of
    control, 55, 59; on poultry houses, 52;
    on poultry market, 18; on women's role
    in poultry farming, 17
Wickard, Claude: on farm exodus, 26–27,
    33–34; World War II and, 23, 28
Wilson and Company, 31
women: cotton farming and, 26; marginal-
    ization of, 3; poultry and, 16–17
Wood, J. H., 22
World War II: broilers, numbers of, 24;
    cotton production, 10; defense indus-
    try, 25; farmers, 25, 33; food rationing,
    xi, 23, 27–29, 35; poultry feed shortage,
    28; poultry production, 23; veterans of,
    34–37

ENVIRONMENTAL HISTORY
AND THE AMERICAN SOUTH

CPSIA information can be obtained
at www.ICGtesting.com
Printed in the USA
LVOW11s1833200617

538756LV00002B/348/P